PATTERNS
IN THE
VOID

D0041277

PATTERNS IN THE VOID

Why Nothing Is Important

STEN F. ODENWALD

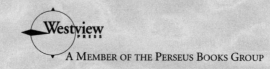

Westview
PRESS
A MEMBER OF THE PERSEUS BOOKS GROUP

Copyright © 2002 by Sten F. Odenwald
Westview Press is a Member of the Perseus Books Group

Library of Congress Cataloging-in-Publication Data
Odenwald, Sten F.
   Patterns in the void : why nothing is important /
Sten F. Odenwald
      p.   cm.
   Includes bibliographical references and index.
   ISBN 0-8133-3938-3
   1. Physics—Philosophy.   2. Nothing (Philosophy).
3. Metaphysics.   I. Title.
QC6 .O35   2002
530'.01—dc21

                                                                 2002004934

Text design by *Brent Wilcox*

First printing, May 2002
1 2 3 4 5 6 7 8 9 10—05 04 03 02

I dedicate this work to my brother Leonard,
in remembrance of all the wonderful moments we had together,
discussing space, time, and the great mysteries of life.
Your presence is missed, but your spirit lives on.

CONTENTS

PROLOGUE

In 1982, I began writing a series of notes to myself that were supposed to summarize how far science had come in understanding the detailed structure of the physical world. I had managed to keep up with the strange discoveries emerging from high-energy physics since I was in grade school. After many years of taking college courses in astronomy and physics, I wanted some way of tying all the loose ends together in my head without embarking on a second Ph.D. in physics. Every problem set I solved for a course or article I read on some exotic new discovery in physics only led to more questions. There were many technical and conceptual questions I wanted to find answers to for my own peace of mind. My graduate-school study in astronomy had been so specialized that I had to put aside keeping up with progress outside my specialty, but now I could return to these subjects with renewed passion. Soon the notes took on the shape of a book because of the many loose ends I kept running into along the way. Exactly what was a "field" as a physical thing? How can a vacuum have structure in it? Why won't Einstein's "fudge factor" go away? My book to myself was called "An Historical Guide to Forces and Fields," or as I often referred to it privately, "Quantum Field Theory for Idiots."

During my lunch hours over the course of the next eight years, I wrote chapter after chapter, probing this wonderful world with grand abandon. Sometimes I would move closer to understanding a fine point in one area. Other times, I felt like I was over my head in mathematical minutia. Yet it was thrilling to read centuries-old ideas and see how they flowed into the modern work of grand unification and string theory. Eventually I began to see that there was a much deeper story behind what I was writing. I had begun my journey as an astronomer, discovering the many things that occupied space, but now I found to my astonishment that it wasn't matter that "mattered." Behind it all, and seemingly aloof from any consort with the things we can see, there is an invisible world that seemed impossibly difficult to describe or assimilate into a comfortable view of the world. It seemed to me that the more that physicists tried to explain this invisible world, the more they got lost down some technical corridor. The final straw came when I stumbled upon a tantalizing quote by Einstein, "Space-time does not claim existence on its own but only as a structural quality of the field." This perspective seemed to be the undercurrent beneath most of the advanced ideas I kept encountering, especially in the area of physics having to do with what has been called the "theory of everything." And then came the discoveries out of my own field that heralded the existence of "dark matter" and "dark energy," the nearly indisputable proof that our universe will spend most of eternity as a starless, dark "thing" devoid of life and consciousness. In a few years, my entire outlook on the physical world had changed in a striking way as physics and astronomy ventured in a different direction in the 1990s. It wasn't stars and atoms that determined the cosmic future. It was space itself that controlled every nuance of what existed in today's universe, and what would exist in the future.

These thoughts would not have troubled anyone else in my profession, but in me they found a curious resonance. It was a coming together of personal experiences and interests that indelibly colored this subject in my mind. I have always been interested in brain research and perception. The many books I had read over the years had given me a rough idea about how the senses worked, how their information became integrated in the brain via complex synaptic networks, and how the mind created maps of the world to navigate. As a child, I was also well taught by my Swedish mother that there really were such things as spirits and ghosts. Like many people in the 1960s and early 1970s, I read intriguing stories about "ancient astronauts" and the mystery of Atlantis and the Bermuda Triangle. UFOs were always mysterious, and even ESP was becoming a fashionable area of investigation at some major universities. Although I have never seen a ghost or had a telepathic experience, long before my outlook had settled upon the scientific way of exploring the world, I had gone through my entire childhood with a rather malleable perspective toward the physical world. Brain research told me that there is more to the world than what perception allows, and especially that things are sometimes not as they appear. My contact with physics as an adult only strengthened these ideas as I explored the half-real worlds of quantum physics and watched reality itself being ripped apart by quantum strings. But Einstein's quote was an even more disturbing thought to wrestle with because it refuted, once and for all, the notion that space and time are preexisting features of our world. Of course I had known this through my technical studies of general relativity, but until 1993 I had never taken the foundations of general relativity to heart. I had compartmentalized its meaning elsewhere in my mind as I raced along studying its mathematical elegance.

I realized, as I reached the end of my journey in writing my private notes, that there was an even bigger story I needed to tackle, now that so many historical threads and personal concerns had been laid bare. During much of the 1990s I tried various approaches to rewriting the older book. It had served as a kind of pencil sharpener that allowed me to hone in on the larger issues I was trying to formulate and understand: Exactly what is space? How does it manage to control matter? What is it about space that forces the world to be three-dimensional? I retitled my work "The Accidental Vacuum," because that title seemed to be closer to the spirit of what I wanted to think about. I wrote two complete manuscript drafts before I finally walked away from the whole project as a hopeless task. It was too technical a story to tell the nonscientist. There was too much that the reader needed to know to appreciate why I thought there was an issue in the first place. Heck, my own colleagues couldn't understand the point I was trying to make or why it seemed to bother me so much. So what if space and time were part of the gravitational field, or that the vacuum is filled with invisible particles that scurry to and fro? Then, on January 15,

1999, I was reading a book about the ancient Incas, their mythologies, and their terrible solution to avoiding the end of the world. In an instant, I saw my way through to a story that was personally gripping, one that captured the essence of why I was so unhinged by what modern physics and astronomy were trying to say. Here's the gist of the problem as best I can explain it:

I sit in my chair and watch the setting sun play its many shadow games with the distant wall. The light pours through chinks between the leaves on an oak tree in the front lawn and filters through the front window and the lacework curtain. The shadows on the darkening wall dance and quiver. In the patches of light, I can still see the faint patterns of the lace. In the shifting of light and dark, I can see the shapes of the leaves, the optical defects of the windowpanes, and images of the curtain lace. The patterns fade to black around the edges and flow into the encroaching darkness that gathers around me. For an instant I realize that the story I want to tell is not about the things of the light: the atoms, stars, and forces that make up this world. I want to explore the patterns hidden in darkness that compel us to fear the night and be anxious about unseen things. It is not a baseless fear. Most of the universe is in an almost palpable dark form that controls the destiny of matter. The underpinnings of reality are in actuality a spider's web of energy that suspends us above a yawning chasm of true nothingness, and nonexistence. It does so from moment to moment without missing a beat. We live in a cosmos whose evolution and fate is determined by things we can never see. We do, indeed, have something to be fearful about as we look up at the dark countenance of the starless, celestial Void. As we look forward into the bleak, dark eternity to come, we must also confront another disturbing fact: We live in a cosmos that seems unimaginably tuned to making life and consciousness possible, but at the same time, it is also a cosmos that has not allowed for life to be an eternal feature of its fabric.

This is what was bothering me. This is the issue that had been gnawing at me in the back of my mind for the last fifteen years. And no one I knew was really addressing this basic issue. It was a liberating thought because now I would throw away all the technical preambles I had spent so much time obsessing about as I groped my way to this bigger issue. I would let other authors tell the technical story of how physicists discovered "nothingness." That was not the story that had so burned in my mind these many years without my really being able to describe it in words. It seemed that the entire theme of physical science was driving us down a very narrow road of logic, to a conclusion that I instinctively felt like struggling against.

My challenge is to find an anchor for myself within this new world of dark energies and patterns lurking in the Void; to find some point to it all as I stretch my mind to encompass the future history of the universe. Many people can find such an anchor in ancient books and writings through the application of faith. That is not a path I have chosen after a lifetime of being suspicious of how malleable Truth can be. This book is about the struggle by one astronomer to find such an anchor amid our changing knowledge of space, time, and solidity.

For more information about this topic, and for updates on its content, please visit The Astronomy Cafe on the World Wide Web at http://www.theastronomycafe.net. I will be posting material at this site that did not make it into the book. I will also present a host of new images that illustrate key issues in this fascinating story of how science has unlocked the mysteries of the Void.

ACKNOWLEDGMENTS

There are so many people that have contributed to my writing this book that it is hard to know where to start in my thanks. My wife, Susan, has always been there for me as I have stumbled from one idea to the next over the last twenty years. Endless conversations spread out over the years have given me a deep understanding of another point of view, another compelling perspective on life, death, and transformation. My sister-in-law Eileen Rattner, herself a scientist, has always encouraged me not to live in the obvious world of my science but to now and then walk on the wild side of alternative views of how the world might be put together. Birth and death are transforming events. My daughters, Emily Rosa and Stacia Elise, born in 1991 and 1994, confront me with a future of optimism and a tangible, humbling example of people who will survive me. They are the explorers of a future I will never know, of experiences I can only imagine. Just as I obtain strength from those in my family who are still living, I have also found strength in the words and memories of those who are no longer with me. My sister-in-law Darlene died in 1978. My parents, Rosa and Sten, died in 1987 and 1991. My brother Leonard died in 2000. Through countless conversations over a lifetime, they helped me see many different facets of the spiritual world, at least through their eyes. They were all firm believers in a greater reality that was with them even to the end of their days. I find this conviction as much of a mystery today as I did when they were still with me. In time I will try to tell my own children the "Story of Life" from my own perspective. I only hope I can give my own children the same rich sense of mystery and reverence about the cosmos that my own family gave me.

Nearly fifteen years of searching for clarity in my writing culminated in my great good fortune to find Holly Hodder, my editor at Westview Press. She was able to steer me around the shoals of irrelevant scientific and historical minutia and helped me get on with the telling of a much deeper story. My copy editor, Michele Wynn, did a masterful job of making the manuscript consistent with the English language.

I have spent decades reading the various popularizations of physics and astronomy that have been written by some of the major figures in these fields, but they generally leave me unsatisfied. They are masterful, technical road maps of the physical world, but they sidestep the more interesting issue of how the writer feels about it all. Among those works I have read, the ones that come the closest to touching upon the cosmos as I have come to know it are: Frank Wilczek and Betsy Devine, *Longing for the Harmonies;* George Smoot and Keay Davidson, *Wrinkles in Time;* Heinz Pagels, *Perfect Symmetry;* and Brian Greene, *The Elegant Universe.* A whole universe of books about brain research and how our minds fashion sense from stimuli has also emerged during this time: John Eccles, *The Human Mystery;* Judith Hooper and Dick Teresi, *The Three-Pound Universe;* and Morton Hunt, *The Universe Within.* The most recent additions to this list are V. S. Ramachandran and Sandra Blakeslee's *Phantoms in the Brain* and Richard Cytowic's *The Man Who Tasted Shapes.* All of them have given me the grounding I needed to appreciate the complexity of brain physiology and the many startling discoveries that have emerged from this research in the last few decades.

Among other written works that have influenced me, I cannot omit a genre that has been a constant during times of joy and sorrow, and during all the other times of my life as well. Science fiction has been my constant companion because of the hopeful and optimistic universes these writers invent. The stories are too numerous to list, but the special ones that held my wonder across the years have been Roger MacBride Allen, *The Ring of Charon;* Isaac Asimov, *The End of Eternity;* J. Ballard, *The Voices of Time;* Stephen Baxter, *Ring;* Greg

Bear, *Legacy* and *Eon;* Gregory Benford, *Cosm;* James Blish, *Cities in Flight;* Arthur C. Clarke, *The Songs of Distant Earth* and *Rama;* John Cramer, *Einstein's Bridge;* C. S. Lewis, *Out of the Silent Planet;* Jack McDevitt, *The Engines of God;* Thomas McDonough, *The Architects of Hyperspace;* Michael Moorcock, *The Sundered Worlds;* Michael Reaves and Steve Perry, *Hellstar;* Carl Sagan, *Contact;* and A. E. van Vogt, *The Wizard of Linn.*

The greatest inspirations I experience often do not come from the written or spoken word. A part of me yearns for nonverbal patterns to ponder in time and space. I am indebted to the wonderful moments I have experienced listening to the music of many artists. I can't tell you exactly what role they played in my thinking about various parts of this book. Sometimes it wasn't even the lyrics themselves that were important but the texture of the sounds that they wove in time at the specific moment I heard them. For me, they helped stimulate my thoughts and feelings in many complex ways. Eric Satie, "Gymnopédies"; Debussy, "Prelude to the Afternoon of a Faun"; Madonna, "Nothing Really Matters"; Enya, "Orinoco Flow"; Emerson, Lake, and Palmer, "From the Beginning"; Sting, "Upon the Golden Fields" and "Synchronicity II"; Cindy Lauper, "Time After Time"; Simon and Garfunkle, "So Long, Frank Lloyd Wright"; The Doors, "You're Lost Little Girl"; Bryan Adams, "Run to You"; Michael Jackson, "Human Nature"; Little River Band, "Reminiscing"; Gerry Rafferty, "Baker Street"; Tomita, "The Sea Named Solaris"; Tim Weisberg, "Dion Blue"; Earth, Wind, and Fire, "Fantasy"; Carly Simon, "Legend in Your Own Time"; Judy Collins, "Both Sides Now."

I am deeply thankful to all of these family members, friends, writers, and artists who have helped this scientist to explore one of the most mysterious facets of existence and to fashion from all their inspirations an outlook on nature that continues to unfold in richness and mystery with each rising sun.

1

THE DARK CONSTELLATIONS
Why We Fear the Dark

In space . . . no one can hear you scream.
—*Alien* movie advertisement

The Fifth Age of the Inca Empire was an age the shamans said would end in a flood that would engulf the universe. It would be the final end to the world, a fate sealed when the Fifth Sun was quenched by the waters of the Milky Way's celestial river. To hold back these waters, Yucana the Celestial Llama stood alone in its dark silhouette against the stars, but there was no telling how successful the Great Llama might be. The price of failure was enormous. It literally involved the end of time itself. So the Incas did the only thing that made sense. They created a powerful system of rituals to help 6 million people assist the Celestial Llama in its labors of holding back the floodwaters and to plead with Wiraquocha to spare them this miserable fate. The 328 tribes of the empire erected 328 Sun Pillars in a circle centered on Cuzco, each honoring the Sun on a specific day, presided over by its own priest.

She was a small child barely ten years old, with long black tresses braided into a flurry of tight dreadlocks. Perhaps her parents had offered her to her tribe's shaman. Or perhaps she had been selected by some complex lottery in her village; the details are lost to us. All who knew her were awed by the honor that had befallen her. For perhaps a month, a year, or even her entire life up until that time, she knew she was destined for greatness. The priests told her this many times in their intimidating and solemn way. Her parents may have been proud or even humbled that one of their own clan would soon be making a holy pilgrimage. Her time was coming in a few weeks. Her gown was being readied, along with delicately made slippers and a warm cloak to cover her body in case of the cold. On her selected day, when the Sun rose over her village's Sun Pillar, she joined a procession and walked or was carried by bearers from the center of Cuzco to an ultimate destination in her own village. It was a month-long journey steeped in religious ceremony, and she passed through strange forests until a familiar outcrop of rock or perhaps a patch of trees caught her eye. Soon she began to recognize her village surroundings. Her pulse quickened. Her heart fluttered with anticipation. In a few hours the procession, chanting all the way, arrived in the village center at the local shrine for the Sun. That evening, she consumed sacred coca leaves. The priests joined her in this ritual of drugs and alcoholic drinks. She had never tasted such things before. She was soon stupefied as the potent chemicals entered her young system and took hold of her. As the Sun began to rise, all the villagers and her parents looked on. The chanting rose in volume as the twilight gave way to the Sun's first rays: Wiraquocha had awakened. A raised hand plunged an obsidian blade deep into her chest. With a few quick strokes, her beating heart was dislodged long before it would ever know love.

Her spirit was released to carry a message to Wiraquocha to pass over the Inca Empire one more year. Each day, a different girl in a different village was dispatched. Her voice added to the hundreds of others each year to create an unending plea to save their world. The priests would also visit every shrine along the way, depositing a tribute of gold, food, and blood from the thousands of llamas they ritually killed each year. The hope was that the voices raised by the multitudes of child messengers would find their way to the ears of Wiraquocha. They would compel him to intercede in the destruction of the Fifth Sun at least for this year. Since they had first decided on this horrifying course of action, the empire had not been disappointed. For over 100 years, the sacrifices had worked every day to guarantee that tomorrow's Sun would rise again.

A big part of me finds their worries silly or at best unfathomable. How in the world can anyone look at the starry night sky and feel afraid of anything? Most of us just stand and stare at the heavens, letting our minds get sucked into the immensity of it all. The Incas were not alone in their worship of the heavens, though they certainly viewed what they saw there with a greater sense of anxiety than most other cultures. It seems humans have always marveled in awe at the mystery of the night sky. We have wondered about the many odd things that have been seen there: wandering planets, meteors and comets among them. As you watch the night sky with its glorious carpet of stars, at first earthly thoughts rush about like a clanging background noise. Then you enter an almost wordless state. The astronomer stops hearing an inner voice lecturing to him about the cosmos. The eye picks out patterns and sees that the vista is far from an empty blackness. How strange it is that we have never failed to invent patterns in the stars to honor the heroic figures of our legends. Chinese, Egyptian, Arabic, Babylonian, Greek, Roman, and American Indian—the roll call of celestial artisans spans the millennia.

In South America, the Incas paid homage to different patterns, not those outlined by our familiar pinpoint stars. Instead, they were bounded by the yawning gaps of darkness between. Speckled along the Milky Way, the dark constellations of Yutu the Bird, Hanpatu the Toad, and Machacuay the Serpent wind their way through the star clouds. The Celestial Llama, Yucana, is a dark rift slashed across the Milky Way as it passes overhead across the Inca Empire. It was Yucana that held back the cosmic floodwaters destined to drown the Sun for eternity. In this blot of emptiness, the Incas found their darkest thoughts realized. If you look farther back in time to the dawn of civilization and beyond, it's not hard to wonder what thoughts may have crossed the minds of still more ancient ancestors. Eventually, there is another aspect to the starless Void that begins to intrude. It is an aspect beyond the primitive logic of minds not yet formed. It is an aspect beyond mystery. It is very primal and visceral. It is something we can express in a single word: Fear.

It's not hard to imagine why it is that rational human beings should fear the dark. As our ancestors entered caves for shelter or crouched in terror on the killing fields of the African savanna at night, they often found these dark places occupied by hidden prowlers eager to make a meal of them. Any contact with darkness was an instinctively fearful event that probably helped preserve them against their natural predators. Yet you can see on the deepest cave walls in France and elsewhere how our remote ancestors crept down dark corridors to paint their figures. It must have been a fearsome journey into nothingness, ending only in a well-lit cavern with hand-drawn animal shapes playing hide-and-seek on the walls in the flickering light. Collectively, our ancestors hunted five-ton mammoths by daylight as adrenaline coursed through their veins. But when entering a cave, perhaps as part of an initiation ceremony, or dealing with the encroaching nighttime, they had to find some way to make peace as individuals with darkness. It has been an ongoing battle for millennia. It is here that I can appreciate the very personal side of dealing with the dark night sky and the hidden things it may contain. You see, I have a confession to make: I have always been afraid of the dark.

As a child, I could never find the courage to dangle my feet off the side of the bed at night. When I entered a darkened room, my hand would quickly dart for the light switch. It was an instinctive fear, but it wasn't one of hidden monsters lurking in the shadows. For reasons I can't fully describe, it was a fear of darkness itself. To my mind, darkness wasn't just the absence of light. Who really knew for sure? Perhaps it really was filled by the spirits of the dead, or by other "things" that could not be seen. The ancients called this unfilled, empty space the Void and considered it an impossible contradiction. It might be dark and free of substance, but there was still some invisible essence within it. All I knew was that its cloaking, smothering embrace could be chillingly sensed, and that seemed to give it an almost palpable presence. As the years passed, it became easier to roam the house at night without turning on a single light. I allowed myself to think that many of these silly ideas had at last been put aside now that I was an adult. But now and then, I would be rudely reminded that I had not quite vanquished my old fears. On a moonless night at an observatory, I was preparing for an evening's research only to realize I had left an important notebook in the trunk of my rental car. With no flashlight and in total darkness, I was forced to walk the 100 yards between observatory and car, which filled me with a mounting sense of fear and panic. Adulthood had done little to convince me that darkness was a benign neutral state. It could still scare the wits out of me under the right conditions, and you will still never find me roaming the basement or attic in the dark.

It is a bit odd that someone with a twinge of fear of dark spaces could have been so compelled to become an astronomer. I find it a very curious attraction, not unlike a moth's fascination with a candle flame. I have spent much of my lifetime investigating the things that populate space—planets, stars, and galaxies—but none of these fill me with a sense of awe or make me in any way anxious or fearful. No, the things in the universe that I cannot help but react to viscerally are those great, dark emptinesses that swallow up everything light-year after light-year. A dark nebula in Orion, glimpsed in Plate 1, looks like some mysterious, shadowy humanoid. During the daytime, I look up and see a comforting blue sky and fleecy clouds. By night, the pinpoint stars in the firmament that catch my eye are all but swallowed up by vast, black, empty expanses marking the depths of interstellar space. For generations, the night sky has been humanity's porthole into the cosmic, the mysterious, the transcendental. For countless millennia we have tried in vain to paint a human face upon infinity. It has existed throughout history as a haunting reminder that some things have remained almost completely beyond our understanding, no matter how hard we have tried to dress them up in the trappings of familiar symbols and terms.

Just as it is hard for me to admit my instinctive fear of darkness, it is equally difficult for me to accept that after all these years of professional study, I know virtually nothing about space. What I have learned over the years about the physical world has done little to comfort me as darkness wraps its cloak about me in my room. It certainly hasn't done very much to quiet my childhood fears that there really may be things that go bump in the night and that most of our world may not be all that it seems. It is a fear beyond logic, and perhaps it is something you can never eliminate completely by talking yourself out of its emotional grip. What really disturbs me as an adult is that the research that astronomers engaged in when I was a fearful child has now turned with alarming speed to focus on the unseeable, hidden side of the cosmos. Astronomers often joke about the dark matter and dark energy they uncover in the intergalactic void. We mockingly refer to these as the "dark side" of the universe as though they are something out of a *Star Wars* episode. But there is a deathly serious side to the study of space and the mysterious, starless Void that has now become impossible to dismiss professionally. We can't dismiss it as individual scientists grappling with their own personal fears, nor does it seem that we can dismiss it as a community of minds probing the frontiers of the physical world. No matter where you turn in physics and astronomy, you eventually find yourself face-to-face with otherworldly hobgoblins that refuse to go away. Only now, instead of simply "darkness" or "Void," we hang other names on them such as "space" or "space-time" or "vacuum" and hope that by assigning clever names to the beasts, we can tame them, somehow. Like the enveloping darkness that unnerves us, they seem to evoke feelings of awe and mystery that for some of us border on the supernatural. There is something deeply unsettling here, something scientists seem always on the very threshold of understanding, but like a pair of ancient sprites, Void and Space constantly race beyond, deftly evading full comprehension.

Space enters our perceptible world only in an oblique way. Because of this, we have to look carefully into our daily experiences to remind ourselves that there is something there to wonder about. You need only look as far as the page of this book you are now reading to experience one of the most ancient and puzzling mysteries of the Void. You see the page and its letters; you do not, however, see the space that separates the page from your eyes. Physics and physiology guarantee that you are intimately aware of the endpoints of a light ray's journey, between the marks on the paper and the reflected light striking the retina. You are never aware of the journey itself. Light brings you information about the world and tells you wonderful things about the surface from which it was emitted: The delicate colors of a rainbow, the texture of a loved one's cheek. But you only experience space in a rather accidental fashion. It is the surfaces that catch your attention: the walls of the room, the texture and form of the sofa, and so on. The underlying space that things occupy is hidden as though it were located in some kind of blind spot. Only by carefully probing the things we see can the contours of the invisible space they inhabit be brought out clearly. To study space as a thing in itself, we have to learn how to look at Nature out of the corner of our mental eye.

Physicists have shown us that at least what we humans call space is something that extends all the way down to atomic scale and even inside the atom. Without this continuity, atomic particles would be scrunched together into a single point with infinite density. At the other extreme, astronomers have surveyed the extent of the visible universe and see space continuing for light-years on end, to larger and larger scales of distance above our heads. All of these explorations return a consistent pattern, a consistent mutually supporting network of data that convinces us that dimensional space is real, that it is a seamless stage for matter's performance, and that it has an objective reality apart from our individual minds and senses. Is it really empty? Most of our ancestors didn't think so. To them, "outer space" was filled by a Fifth Essence, along with different kinds of crystalline spheres to keep the planets in their courses. And why not? Today we can stand outside at night and think about outer space. It is cluttered to varying degrees with stars and stray atoms. It is certainly anything but empty. Even the ancient Egyptians 4,000 years ago had it right. The entire world was filled by the body of the goddess Nut, whose substance was air and whose likeness in Figure 1.1 graced many tomb ceilings. On a moonless night in the warm evening deserts of ancient Egypt, the darkness of the ground blends with the darkness of the sky in a seamless way. The illusion is complete. With every breath you take, the Void is vanquished.

There is also another reason the Void can't be completely barren. Gravity pins me to the ground and forces planets to orbit their stars. How could gravity work in a true void if there wasn't something al-

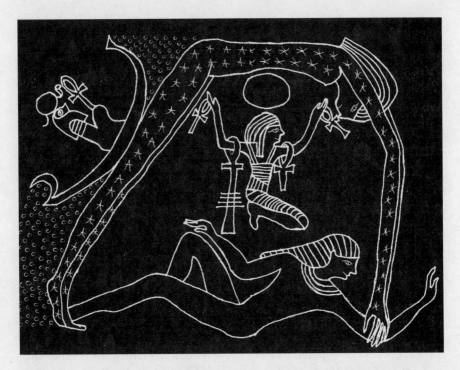

FIGURE 1.1   The sky goddess Nut spreads her star-filled body across space and banishes emptiness. The sun god Ra rides a solar boat across her body to separate day from night. Shu, the god of the space between Earth and sky that is filled by the goddess Tefnut (the moist atmosphere) supports Nut. Behind it all is Amen, the ancient supreme god of the Egyptian pantheon, whose name literally means "what is not seen" or "what is hidden." In his incarnations as Amen-Ra and Kheperu-Ra, he is the personification of the hidden and unknown power that was associated with the creation of the universe, including Nut, Shu, and all the other gods and goddesses.

ready there to transport its pulls? In the hands of twentieth-century science, gravity became something physical, a nearly incomprehensible "thing" that had in its fabric the properties of both space and time in an impossible webwork of contrasting qualities. The relationship was so intimate that if you could shut off gravity, both space and time would actually vanish. Then came the equally incredible discoveries about the nature of matter and the other forces that ply the Void. To make the discoveries and theories understandable, the Void had to be turned into something far more complicated than simply empty space. Darkness had a shape to it that could literally steer the stars while remaining utterly invisible.

As scientists became better at understanding how matter and force played out their complex choreography in the atom-sized dimensions of space and time, they also found out that the Void is filled with other things, too. Physicists began to detect ghostly particles of matter existing beyond the veil of direct observation—a hidden companion universe to our own, filling space with effervescent activity and causing matter to jiggle imperceptibly. Matter, they thought, could be affected by this hidden world in very complex ways, even though no amount of scrutiny could ever show us the details of how these influences worked. They found that fields of energy would shimmer and dance, under a cloak of complete invisibility because Nature forbade any direct observation of them. Like Medusa, the Void could only be watched indirectly through the various mathematical mirrors that we, like Perseus, might fashion. One thing was certain, there didn't seem to be any limit to how deeply this hidden world within the Void could entangle itself with our own. As physicists created ever more accurate descriptions of the matter and the forces they could see directly, new patterns in the Void had to be added along the way to make the explanations work. These invisible patterns allowed matter to have heft. They allowed forces to have a distant reach through space. But they did something else, too. Darkness had now acquired both form and substance.

At the other extreme, astronomers discovered that cosmic space is vast beyond human comprehension, and its properties seem to be no less enigmatic than what we find at the atomic scale. For many civilizations, the primordial, cosmic Void was a state of pure space, though within its dark shroud, it hosted vague motions and potentialities for bringing into existence the physical world. Creation legends from around the world tell us that all that was needed to bring forth a solid world was a divine stimulus acting on the Void. As we have learned how the Void works, we have also become the latter-day Promethians and have again stolen fire from the gods. Physicists routinely pull matter out of the Void in their laboratories. From these labors, astronomers have begun to understand how our universe may have been born. Each time we slam particles together in our labs, we shake the fabric of space itself, and out of its darkness tumble forth myriad new particles. What we do in the human-scale arena is now seen as only a brief glimpse of more cosmic engines of creation that worked their magic billions of years ago. Out of the titanic changes that attended the creation of the universe, the Void trembled. A hailstorm of matter was torn out of its fabric. An incomprehensible burst of annihilation converted the darkness into light, leaving behind in its wake the merest trace of matter. This matter eventually became galaxies, stars—and ourselves.

I think one of the most unique moments in the history of science came when physicists realized that space is the physical embodiment of gravity. This "empty" thing that we see out of the corner of our eyes, or in the blackness of the night sky, is not a passive receptacle for matter and energy. Cosmic space can be bent or warped by stars and distant galaxies. This bending action gives the universe a definite shape in time and space, and from this combination, a unique future. But the shape of the cosmic Void is only partially controlled by the stars and galaxies we can see. There is a much larger ingredient of the cosmos that refuses to show itself directly. We are forced to confront the fact that something hidden in the Void is controlling not just the subtle properties of matter but the destiny of the universe. And thanks to the machinations of this invisible energy, the death of our universe will be a bleak, starless darkness propelled into unending eternity.

So, whether it is atomic or cosmic, the Void is a mysterious empty landscape that continues paradoxically to be rich in hidden detail at every scale, with properties that never fail to surprise us at every turn. Behind it all, only gravity seems to be the key to understanding what is really going on. It is here that we learn another secret about space. In the end, the Void will take back all of the riches it bestowed on matter at the creation of the universe and will replace it with a desolate, ever-expanding space, leaving not so much as a single star or atom behind.

The Incas were the first to imagine that darkness can be as important as light in defining their destiny. It was a horrible fate that they saw in the Milky Way's dark dust lanes—a consuming emptiness shaped in just the right way, forming its own terrible patterns. In these shapes within the shimmering Milky Way, they saw a complex story of grim earthly events to be played out each year. How anxious they must have felt to see the same dreadful warnings repeat themselves each year. The Incas tried desperately to change these events by calling upon the darkest actions of the human spirit. They created an elaborate ritual and ruse that spanned a century or more. Its singular goal: to force the universe to favor the continuance of their existence through the worst kind of sacrifices. And it worked. Their cosmological beliefs led them down an increasingly terrifying corridor of logical deduction, but their yearly rituals offered a hopeful way out, an expensive escape clause they could invoke year after year to indefinitely postpone the end of the cosmos.

Today we find ourselves confronting patterns in the same dark abyss that worried the Incas, only to discover an equally troubling landscape. We study dark constellations in other galaxies, as shown in Plate 2, and wonder what lies beyond. In many ways, this modern vista is more disturbing than any vision that an Inca priest could have imagined, for now it is the dark energies hidden in the Void that circumscribe our destiny. In the Void, we see shapes, and hidden movement that in the end will seal the fate of our own space. Some fear that not even the atoms in our own bodies will remain to bear witness to the story of our passing. Only the Void itself will outlast us. There is no escape clause or ritual we can invoke to delay this ending. We can only regard it with a sense of remote fascination. A few among us may even view it with a sense of philosophical despair. How dreadful that the universe we see today will eventually just wink out, with only an infernal darkness existing in the end. And just like our ancestors, we find ourselves having to make peace with the Void, and its encompassing darkness.

So we are left to study the things that we can and to search among the flotsam and jetsam of the physical world for a key to understanding what it all means. Eventually, we find out that if the Void is a door we must open to understand the deep meaning behind our existence and our fate, then gravity is the key to unlocking its mysteries. As Einstein taught us nearly 100 years ago, gravity is both time and space. It is the creator of planets and stars. It shapes the universe and bends the fate of matter into a collective destiny.

The struggle I have had nearly all my scientific life is to really accept that gravity is so terribly important in actually sculpting the Void. How does it demand that space must be counted by only three directions or that a ruler can't be turned through other dimensions? How is it that the matter we see becomes a slave to something that is invisible? Is matter like the raisins in bread, connected but still separate? Does matter blend seamlessly into gravity like the threads in a tapestry? Somewhere among the answers to questions like these are the clues to explaining how invisible gravity can create time and space or light a rainbow with color. But what is it about gravity that makes this possible? I long ago realized that if I wanted to search for the answers to questions like these, I would have to figure out a way of overcoming a basic experimental problem: How in the world would you study something like gravity or space when you can't touch or weigh them? My education in science provided me with a simple answer to this paradoxical question: Sometimes the best strategy isn't the direct approach at all. I could stare into the night for millennia, but I would never make much progress. Suppose, instead, I tried an indirect approach. Some problems can be solved by literally sneaking up on them by solving other related problems.

Imagine yourself on an island in the middle of the ocean. Having nothing better to do with your free time while waiting to be rescued, you decide that you might as well learn something about the ocean that surrounds you. So, you walk down to the seashore, and you study the tide pools. In this transition zone between water and land, you soon learn about the regularity of the tides, the chemistry of the water, and the ecology of the organisms that are nurtured by the ocean. In a similar way, we can study the invisible space where we live, by studying where it laps up against the physical bodies we see in space, from quarks to quasars.

Our scientific exploration of the physical world through frequent glances into its many tide pools has indeed been undeservedly profitable. Physicists have uncovered a richness to the physical world scarcely imagined a century ago. Among the many particles and fields that are embedded in the Void, a bewildering patina of cause-and-effect patterns interlocks across space and time to create law and order. Some of these patterns are hidden in places where Nature does not let us see. In the borderland between what is seen and what is not, science at last seems to be making a tenuous contact with the intangible essences of things that are more like spirits and ghosts than the compelling firmness of a solid world. Just as we begin to catch glimpses of a larger world beyond the tide pools, we find ourselves suddenly lost in a completely alien landscape without any familiar anchors. Without the reassuring shapes and reference points of solid matter, how are we to understand the shifting patina of the fields that underlie them? If the dross matter that forms our bodies merely consists of waves of energy crystallized for a moment in time, what new ideas do we have to entertain to connect ourselves with the larger and more insubstantial universe?

We also need to make peace with the miracle of our own existence, with how it hangs suspended between the darkness of the Void and the fleeting brilliance of the stellar universe. In the end, we have to create, as our ancestors did, a new story for the modern age—so we will no longer fear the Void, or the darkness that it brings, any more than we fear our right hand or the color of our hair. As we enter the twenty-first century, the scientific investigation of space has begun to show space as an almost incomprehensibly alien landscape in which the rest of existence is embedded. In the end, space ranks as the greatest enigma that physical science has yet to confront. It literally prevents reality from collapsing into nothingness. It is the origin of time, mass, energy, and as far as we can tell, even the universe itself. Both the miraculous story of our origins and a bittersweet glimpse of our destiny are written within its dark tapestry. The recognition that this is so has taken centuries to come into its own. In a matter of a few short decades, it now washes over nearly all areas of physics and astronomy. What I now want to share with you is the story of how this particular revelation came into focus for me, and what I think are its personal ramifications.

I want to start by reflecting upon the struggles of our ancestors as they tried to understand how a void could be filled, yet seem empty. Despite the thousands of people who have thought about the Void, only a handful of ideas came out of all this activity prior to the twentieth century. I want to guide you through encounters with the dramatic insights into space, time, and field provided by twentieth-century physics. I won't pretend that this will not be tough going, but as you will see, they are the breakthroughs that have pushed the exploration of space into areas of thought beyond our wildest imaginings. The reason we have been so unsuccessful in understanding the Void is because it requires the mastery of a new logic, which we had to uncover in the atomic domain. We will also see in subsequent chapters why modern astronomy has been turned upside down by the discovery of dark matter and dark energy. Finally, I am going to touch upon the mysterious nature of the Void as seen through the eyes of contemporary physicists and cosmologists. As we explore these details and patterns that form the bedrock of the physical world, we will try to create answers to the questions that seem to arise over and over again. How do we find ourselves within this spider's web of shimmering fields and things that go bump in the night? How do we stand up to darkness and confront it without flinching or running away from it in terror? Where among the jumble of fields that make up our own physical bodies is there room for a hint of a spiritual or conscious essence? The best way I know to find clues to the answers is to examine my personal experiences of mysterious events in my life and see how they have helped me understand intuitively what these scientific discoveries mean. The next chapter will explore the first attempts at filling the Void and will explain how some ridiculously simple experiments can put you directly in touch with things that really do go bump in the night.

2

THE SPIRITS WITHIN
Invisible Fields and Ethers

The frank realization that physical science is concerned with a world of shadows is one of the most significant of recent advances.
—Arthur Eddington, 1929

Sometimes when I wake up in the middle of the night, I keep my eyes closed and listen to the noises that my body makes. I can hear my heart beating, or at times the crackle of a noisy joint in my neck. If I want to, I can even tense my facial muscles and hear their rumbling drone. One sound in particular has always been with me and is less manageable than the others, less willing to perform at my command. It is the electrical hiss of my tinnitus. Severe childhood earaches left me with this all-too-common ringing in my ears. There is a distinct, high-pitched graininess to it that reminds me of a television set tuned between channels. A part of me understands the medical reason for this sound, but another part of me imagines my mind trying to tune in to another world waiting for some message to arrive. None ever does. At times it can be annoyingly loud, especially if I have a fever or have had too much caffeine, but most of the time my tinnitus plays peek-a-boo from behind the familiar sounds in my life. It can be an unnerving experience to navigate a dark room with only my heart sounds and

15

tinnitus for company. You might think of it as the soundtrack of darkness.

When I get tired of listening to my inner noises, I sometimes apply a gentle pressure to my eyelids to check up on what else is going on inside my body. Slowly at first, but then with increasing tempo and variety, I am treated to a spectacular light show. Again this is a product of my internal world, though the brilliance of the patterns that flow out of the dark seem unnervingly intense. It's a very strange experience to see lines and bursting grids suddenly appear out of the darkness beneath my eyelids. These phosphenes, as they are called, are easily overwhelmed by the light that leaks through my eyelids in the daytime. Only the dark night backdrop gives me a clear impression of these geometric fireworks. As spectacular as they seem, I know they aren't real perceptions of things in the world—they are impostors. The pressure on my eyelids causes some of the neurons in my retina to fire without any benefit of optical stimulation. The brain is fooled into thinking that the retina is seeing something, so it takes the stimuli and creates a model of what it thinks it is seeing. Cells specialized to sense lines and corners announce that they have seen them, and a fabulous carpet of geometric forms is brought into existence. I watch as they appear in waves of activity, timed by some biological clockwork that advances the scenes one pattern at a time.

When my eyes are open in a dark room, I often notice that the shadows and shapes of bedroom furniture are not entirely featureless. There is a distinct graininess that shimmers just above the threshold of perception. Like a pointillistic painting by Seurat, the darkness becomes pixelized in a vibrant, speckled pattern of black and dark gray. Even in the daytime I can see this speckling as I look at a wall or some other uncomplicated surface. Once again, my eye has taken in the outside world, reducing it to the individual firings of retinal cells that leave their discrete characteristics imprinted on my perception of the world. But the graininess is of such a fine quality that I am really not conscious of it at all as I go about my daily chores. Only at night, when there are no distractions, can I watch the grainy shadows. I reflect on how my biology shapes the world that I perceive by regulating the way my mind is allowed to see the world. Between the tinnitus, the magical phosphenes, and the granulated darkness, I have long appreciated how my internal world and my physiology can leave faint but memorable imprints on the external one. For some people under severe stress or autosuggestion, it's easy to understand whence ghosts and other shadowy specters can come. As a child experiencing the same nighttime conditions, I recall vivid impressions of menacing shadows and other fantastic shapes emerging out of the grainy darkness. As an adult, I can see how difficult it will be to understand something as intangible as the Void if I don't first understand how my brain makes sense out of its senses. What does it really mean for me, to say that I understand something? What does it mean to understand the Void?

The brain has evolved to extract patterns out of any stimuli it receives from its sense organs, so it is not surprising that we see constellations in the sky that resemble scorpions or dippers, or dark rifts that resemble llamas. The dividing line between the patterns the brain finds in the outside world and the ones it fashions in our inner world is so tenuous that reality blends with the unreal in a very fluid way. When we explore unfamiliar landscapes and see strange new animals or phenomena, our brains try to make sense of what we are experiencing by referring to other, more conventional patterns as templates. We discount the impossible creatures that a child sees under a bed, but as adults we can also be fooled in more subtle ways. Senior citizens who have suffered from glaucoma, cataracts, macular degeneration, or retinopathy can sometimes experience Charles Bonnet's syndrome, in which inappropriate figures are projected back into the real world. These are not translucent specters, but images that have a genuine physicality to them. The brain actively fills in information in its visual field, replacing unreceptive areas caused by disease with cartoon figures, monkeys, or anything else. As the neuroscientist Dr. Ramachandran at the University of California–Santa Barbara notes in his book *Phantoms in the Brain*, "[I]t's clear that the mind, like Nature, abhors a vacuum and will apparently supply whatever information is required to complete the scene." In most cases, these hallucinations are easy to dismiss because the patient or the observer recognizes they are not real when someone confronts them about what they saw. That patients can be so easily persuaded to dismiss what they saw also suggests that the event was not neurologically connected with the emotion-generating limbic system. This ancient neural pathway inherited from our reptilian ancestors marks every sensory impression with an emotional content, and its headquarters in the brain region lie at the very base of the cerebral cortex. Some hallucinations, however, can get connected with the limbic system, and patients will not dismiss these so readily through any logical probing, because they had a powerful emotional response to the image, which suffuses the memory of the event. Still, the number of people who claim steadfastly that they saw aliens trying to abduct them or the approximately 30 percent of the public that claims to have seen angels might simply be hallucinating. At least in the latter case, because of powerful social prohibitions, no one would think of accusing them of doing so. However, there are other, more insidious visual illusions that are not as easily caught.

Artists in the 1500s who had never seen a whale drew the occasional beached whale with large elephant ears near its head. After all, heads do have ears. The English art historian E. H. Gombrich was aware of how the brain completes missing details in its model of the world but also knew that it doesn't do this without some bias: "The familiar will always remain the likely starting point for the rendering of the unfamiliar." We have to expect that when it is presented with a Void or with dark scenery, the brain will try to extract some understanding from what the senses have registered. This is a matter of survival. The brain will use whatever models, archetypes, or analogies it already has at its disposal and will try to mold the new data into conjunction with older stereotypes. This all connects with our goal of understanding the Void, because to do so we must evaluate sparse data and try to frame an intuitive model or theory to connect the dots. Physicists of the eighteenth century worked very hard to understand the nature of gravity and the various ethers and effluvia that they thought filled the Void. They were forced to do so with little data, but they evidently had many ideas that seemed ready-made to create a plausible accounting of the invisible. Fortunately, when it comes to studying the Void today, we are not entirely forced to use our imaginations to puzzle out a better model. Scientists have discovered that Nature hides clues to understanding the Void within many different systems in the physical world. Sometimes the clues are, literally, hidden right under our noses.

Have you ever wondered what causes a magnet to push iron nails around without any direct physical contact? What is it that causes a bowling ball to drop to the floor when it is released? We call these phenomena "magnetism" and "gravity," but what do these terms really mean? Somewhere between the hard bodies of our world and the empty Void, there is a middle world unseen by humans that enables all motions and interactions. When we play billiards, we see easily how the direct contact between the balls makes them suddenly move. But what is going on with magnetism and gravity? We all agree there is some kind of "influence" or "transaction" taking place in the space between magnet and nail. No matter how much we play with magnets and nails or bowling balls, the agent for causing the forces seems absent from the scene. We can call it gravity or magnetism, and declare in so doing that we know what they are, but these common names actually conceal from us their true nature. The names bring us not one millimeter closer to understanding the nature of the forces themselves. The fact that they are both invisible and yet fill the Void, however, provides us with just the tide pool we need to explore to reach a deeper understanding of how the empty Void connects with the other more solid things with which we are so intimately familiar. Historically, the force of gravity was the first tide pool that anyone tried to explore in detail, so we will begin our study of the Void by looking at this force more closely.

In the history of understanding gravity, few names are quoted more often than that of the English scientist Sir Isaac Newton (1642–1727). His accomplishments are legendary. It was Newton who gave us our first detailed look at how gravity operates, not only on Earth but in space as well. He explained mathematically the motion of cannonballs and planets, and even the ocean tides, thrown in for good measure. If anyone could figure out what gravity was by the modern methods of science, no one by the seventeenth century ever had better credentials than Newton. Amazingly enough, Newton never looked deeply into *why* gravity makes anything move. Why did he stop so short of explaining what gravity's action actually was? As far as anyone can tell from his writings, it simply wasn't a research direction he found profitable. He strenuously resisted the temptation to hypothesize about gravity, not so much because he was lazy but because there was nothing in his mathematics to suggest a plausible mechanism. To him, all you needed to know about gravity and how it plied the Void came packaged in a simple equation known to every student of science: Newton's law of universal gravitation. This, however, was as far as he was able to go. Unlike modern-day scientists who tend to enjoy public displays of speculation, Newton offered no public or formal clue to what he thought gravity was of itself, preferring instead to "frame no hypothesis." Besides, there already existed an explanation for gravity, and it was available to all who felt they needed a more intuitive explanation than Newton's abstruse mathematics.

Some years before Newton began his mathematical study of gravity, the French philosopher René Descartes (1596–1650) hit upon a wonderfully intuitive picture for gravity, which we see in Figure 2.1. If you stand outside on a windy day, the invisible air pushes the leaves on a tree or tugs at your clothing. What could be simpler than to imagine each body buried in its own invisible cloud of particles swirling around it like interplanetary cyclones? These particles, like gnats flocking on a summer's day, would constantly bump into each other to produce the pulls we experience as the force of gravity. Every planet, every mote of matter, would be surrounded by a swirling tornado of invisible particles causing the visible object to move about like a leaf carried by the wind. It was a brilliant idea. You could actually see this activity happening in your mind's eye. However, the idea led absolutely nowhere. Without any mathematics to support the idea, you couldn't use it to make a single prediction about how gravity ought to work—not remotely as well as Newton's calculations, in any event. So, here was the dilemma: If you wanted an explanation for gravity that would satisfy your grandmother, you turned to Descartes. If you wanted to calculate the orbit of Neptune, you turned to Newton.

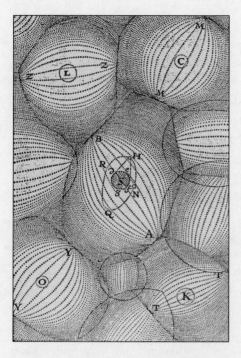

FIGURE 2.1
Descartes's sketch of domains in the Void controlled by various clouds of invisible particles. Invisible vortices of particles engulfed every object and gave space a tessellated but invisible pattern of swirling, force-giving activity. (Courtesy Descartes's *Principia Philosophiae*, originally published in 1644)

Although Newton couldn't be pinned down to a public statement about the nature of gravity, it's not too hard to see where his mind was headed from his letters to friends such as the secretary of the Royal Society, Henry Oldenburg (1617–1677). The letters make it pretty clear that like Descartes before him, Newton also thought gravity was something carried through space by some kind of invisible substance: a substance penetrating to the very core of matter.

> [I]t is to be supposed therein, that there is an aetheral medium much of the same constitution with air, but for rarer, subtler and more strongly elastic . . . For the electric and magnetic effluvia, and the gravitating principle, seem to argue such variety. Perhaps the whole frame of Nature may be nothing but various contextures of some certain aetheral spirits or vapors, condensed as it were by precipitation . . . Thus perhaps may all things be originated from aether.

It is actually disappointing for me to see that Newton's brilliant studies of gravity had led him to an idea that was in most respects no better than Descartes's. But there was a new twist that he added to the older description that is a fascinating new perspective on Descartes's idea. Newton's mental image for gravity wasn't just a simple copy of Descartes's. Newton didn't just wonder whether gravity was created by invisible vapors filling the Void, he also wondered whether *all* of Nature, and *all* things in it, weren't somehow patterns and condensations in this same invisible substance. Here, among Newton's letters, we see a new possibility for what these invisible spirits may imply for the physical world. Could matter itself be simply a *quality* of gravity, condensed like the morning dew on a spider's web? It was an intriguing thought, but like Descartes's invisible particles, there was little that could prove its validity. Besides, the exploration of gravity seemed to have reached an abrupt end as the eighteenth century came to a close. Physicists quietly used Newton's mathematics to calculate orbits for new comets and planets, but left the meaning of gravity to philosophers. There was simply no further data to move the study of gravity's invisible essence further along. But there was another corner of Nature that was being investigated with increasing interest: magnetism. Unlike gravity, you could actually manipulate *this* force in the laboratory.

As you pull iron nails across a table with a magnet or place the same poles of two magnets together, it is impossible not to imagine some kind of invisible emanation at work. Your eyes can't see it, but your fingertips can sense the slippery "something" that forces the poles apart. Not long ago, scientists could only manipulate magnetism by using naturally occurring nuggets of a mineral called lodestone. It took a fair fraction of the nineteenth century before scientists working in their crude laboratories finally figured out how to create artificial magnetism. They used copper wire wrapped with cloth to carry electric currents. University labs boasted about the sizes of their batteries, which often took up whole rooms. Soon they learned how to create magnetic fields that outstripped not only the feeble effects of lodestone but also the much stronger tugs of Earth's own influences.

Sometime in 1852, and inscribed with great care in the last of his mammoth notebooks, the English physicist Michael Faraday

FIGURE 2.2
Magnetic fields are normally invisible to the eye, but their influence can easily be seen in the way matter moves even on the seething surface of the Sun. (Courtesy of NASA/TRACE mission)

(1791–1867) came up with the curious idea that magnetism was carried by "lines of force." He reproduced Descartes's experiments, but saw in them something that Descartes had overlooked. Iron filings sprinkled on a sheet of paper with a magnet underneath revealed beautiful lines of magnetism. We see the same lines in Figure 2.2, but on a much grander scale. To Descartes, they were just a curiosity. To Faraday, they were a glimpse of an invisible world that had a definite pattern. The very thought that there might be invisible patterns, "strains in space" controlling Nature, was instantly liberating. In a great leap of insight, Faraday imagined that all the laws of Nature were the result of the interactions of these hidden patterns in the Void. Like Newton's glimpse of matter as a condensation of gravitational vapors, Faraday had also seen a larger universe of possibilities, this time within the magnetic patterns and strains in space.

What was especially exciting about Faraday's image of geometric patterns etched in space was that they made the connection to mathematics much easier to appreciate in the precise curves they traced out. Scottish physicist James Clerk Maxwell (1831–1879) quickly took these patterns and fashioned them into a theory of magnetism and electricity that was incredibly rich in logical detail and depth. It was in these mathematical deliberations that a new ingredient to the

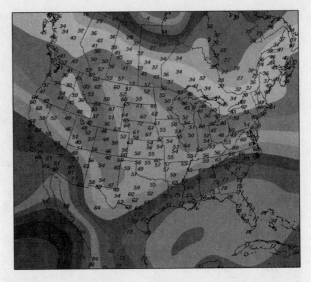

FIGURE 2.3
A simple temperature field taken from an ordinary weather map shows how common certain kinds of fields and their representations have become in our daily lives.

invisible Void came into its own: fields. In the symbols that chased themselves across his pages, Maxwell saw electric and magnetic forces as part of a single natural essence he called the electromagnetic field. To Maxwell's way of thinking, fields weren't just some mathematical abstraction; they actually had a substance to them, perhaps not too different from what Descartes had imagined two centuries earlier.

Science fiction stories not withstanding, fields are actually so common in our world that we take them for granted; for instance, the terms "cornfields" and "wheat fields" describe everyday things. Whenever you measure something at many different points in space, it is called a field. It's as simple as that. For example, in Figure 2.3, you see a common weather map. It shows many different temperatures across the country, and from them you can see that some locations on Earth's surface are warmer than others. To a weather forecaster, this temperature field isn't just something abstract. This field actually causes air to flow, creates high- or low-pressure centers, and drives the weather systems across the globe.

In physics and mathematics, fields can be other kinds of things spread out in space that are a lot less familiar than corn, wheat, or temperature. There are velocity fields, force fields, and quantum fields. There can be fields of charge, fields of density, and fields of magnetism. Some of these fields have an obvious mechanical origin. Each temperature reading on a weather map can be traced to a single person holding a thermometer at a specific geographic spot. Each wind-speed measurement represents a reading made by a person holding an anemometer. When we talk about gravity fields or electromagnetic fields, these are also qualities of the physical world we can measure at different points in space even though, just as for temperature, we cannot see them directly.

The beauty of working with fields is that they can be described and manipulated mathematically. They can be added, subtracted, rotated, flipped inside out, mirrored, and more. When you do this, you can often uncover hidden details you did not originally imagine as properties of these fields just by looking at a reading on a single thermometer. For example, in meteorology, when a temperature field changes abruptly over a short distance, this causes hot air to rise and cold air to sink, creating a local wind that then tries to smooth out the temperature difference. It only takes a simple rearrangement of the mathematical symbol defining Maxwell's electromagnetic field to show that disturbances in this field travel through space at the speed of light and in the form of a wave. The German physicist Heinrich Hertz (1857–1894), already skilled in using Maxwell's mathematics, experimented with generating and detecting these electromagnetic waves. It didn't take him long to see that they traveled at the speed of light and were actually a form of light itself. Not only had Maxwell found a way to describe electricity and magnetism as two complementary aspects of a single field, but he discovered a completely unintended side benefit to this synthesis. He could account for the existence of light itself—one of the oldest wonders of Nature, that until then had received no explanation. These accomplishments didn't come without a price. In shuffling the electric and magnetic cards in Nature's game, a wild card was thrown into this deck. Maxwell had to include a new ingredient in the Void to make the visible part of his theory work out properly. This was some mental scratch paper he was using to set up the equations, and he was convinced that this internal image, like the electromagnetic field, represented something physical. He called it "the Ether."

Maxwell's theory wasn't merely a direct description of electric and magnetic fields. In his mind, it was a description of what the Ether was doing. Ether was the primary object, not the electromagnetic field. Maxwell thought of the Ether, although hidden from view, as something material like water or iron that filled space. He even imagined, as in Figure 2.4, that it penetrated the spaces within every substance. In the end, Maxwell had created a theory for electromagnetism that not only accounted for light but also supported an earlier idea by the Dutch optician Christian Huygens (1629–1695) about light needing the Ether to support its wave motion. The curious thing about the Ether is that it didn't appear as a specific symbol in any of Maxwell's equations. It was thoroughly hidden behind the symbols on the page like some kind of ghostly mental scaffolding provided by the brain to hold up his various mathematical abstractions. It didn't even show up as an enigmatic field of any kind, because without a mathematical symbol to represent it, it simply didn't exist.

As it would turn out, the Ether was our first scientific entanglement with an invisible world that refused to reveal itself through any direct observation. Yet it seemed to be absolutely required so that humans could understand that the symbols in their theories represented something in a fundamental way. When physicists tried to figure out just what this space-filling Ether was, it only led to a host of spectacular contradictions.

If you added up all of the properties the Ether gleaned from a raft of theoretical and experimental findings, the bottom line was that the properties taken as a whole just didn't make sense. The Ether had to be stiff enough to let high-frequency light waves travel through it, but not so stiff that it would resist the movement of the planets around the Sun. The English physicist Oliver Lodge (1851–1940), without embarrassment or apology, estimated that the density of the Ether was 20 billion times the density of solid rock. Yet, if you were calculating the precise motions of the planets, the Ether could be completely ignored. This means it would have to be entirely without any gravitating mass, yet you could still talk about it as though it was a medium with a particular density. But where was it? Despite everything that physicists tried to do to uncover its physical traces, even the best efforts turned up nothing. Oliver Lodge, by the way, was no stranger to exploring the invisible and attempting to quantify it. He was also fond of exploring psychic phenomena.

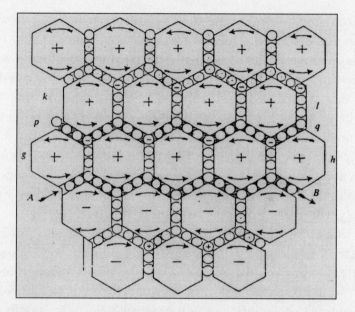

FIGURE 2.4 Maxwell imagined that space was filled with an ether that acted like cogs and gears to move light from place to place. (Courtesy *Philosopy Magazine* Volume 21, page 291)

By the turn of the twentieth century, it didn't matter that no trace of the Ether had ever revealed itself. Nearly everyone still believed that something like the Ether had to exist for electromagnetic waves to travel through the Void. The Ether provided such a potent mental image of what might lurk within the Void that no one was eager to abandon it, especially since there was no better idea to replace it. Some physicists even wondered whether the gravitational fields threading the Void also needed their own Ether to get about. Into this soup of fields and multiplying ethers, ordinary matter would somehow be plopped like bits of beef and potatoes. Then, no sooner had the idea of the Ether established itself in the scientific community than it was rendered irrelevant almost overnight by a new idea in physics: relativity.

One of the most philosophically unsettling aspects of physics at the turn of the twentieth century was that there were actually two separate universes of physical laws. The Maxwellian laws of Nature controlling radio waves in Guglielmo Marconi's (1874–1937) new transcontinental messages appeared to be separate from the Newtonian laws that described planets and cannonballs. What, after all, could the toss of a baseball have to do with the transmission of radio waves? The German physicist Albert Einstein (1879–1955) also wondered about this absurd situation, coming to the conclusion that there could be only one kind of reality beneath it all. He had no physical proof for this conjecture, only an inspired hunch that reality should be *simple*. There shouldn't be two different Voids, two different Ethers, and two different theories to account for events in the world. There should just be one kind of description, but it would have to be precisely tempered by the point of view of the observer.

The title of his 1905 paper that announced these ideas was a barnstormer: "The Electrodynamics of Moving Bodies." For a physicist, this title is as provocative as "How the Color of a Car Determines Its Speed" would be to the rest of us. In one fell swoop, he showed just how Newton's moving bodies and Maxwell's electromagnetic fields were subtly related to each other. The only way to bring these two universes into alignment was to change Newton's physics so that it could handle motion as it approached the speed of light. When this was done, the ways that two observers had to compare notes for what they were seeing flowed together into one description that paralleled key elements in Maxwell's theory.

Imagine that you were moving along with a current of charged particles in a wire, each producing its own electric field. In your frame of reference, the charges are not moving at all and you will see a pure electric field caused by the individual overlapping fields of each electron in the current. Now, if you stepped back into the lab's frame of reference, Einstein's relativity says that you will be in for a surprise. The electric fields are now gone and they have been replaced by a magnetic field surrounding the current flow. In fact, by changing your speed relative to the current, you can also increase or decrease the strength of this magnetic field. Magnetic fields do not exist independently of the relative motion of the observer. What is especially amazing about this sudden change in scientific perspective is that it didn't come from physicists studying electromagnetism in their labs. It came from a dramatically successful attempt to combine the universe of Maxwell and the universe of Newton. The most immediate outcome of this radical change was that the Ether was no longer needed. It wasn't that the Ether idea was finally disproven by a single key experiment, though several did contribute to its downfall. The Ether idea was orphaned as its theoretical underpinnings were finally removed.

Even after Einstein's theoretical breakthrough, and the many negative experimental searches for the Ether, many physicists at the time refused to stop thinking about the Ether. It was as though physicists had dug a mental rut so deep there was no longer a way to think in any other way. To them, the Ether was still a helpful mental image for thinking about fields and waves in space. As Gombrich noted about artists, "[A]n existing representation will always exert its spell over the artist even while he strives to record the truth." Physics textbooks as late as the 1930s continued to view the electromagnetic world through the filter of this peculiar medium. By 1941, English physicist Sir James Jeans (1877–1946) finally blew the whistle on just how absurd the Ether idea had become. He didn't have to offer as evidence a lengthy mathematical or experimental refutation of the Ether. All Jeans offered was a very simple but compelling plea to the physics community that was equivalent to announcing, "The Emperor has no clothes." Was Nature really conspiring in complex ways to hide the existence of the Ether from us in our experiments, or could it be that there never was an Ether in the first place? As Jeans put it, "If we accept this view, there is no conspiracy of concealment for the simple reason that there is no longer anything to conceal."

Like some slowly sinking juggernaut, the Ether idea had outlived its usefulness, and it only survived by force of habit. It was now time to abandon it and get on with more interesting avenues of research. Physicists eventually did adjust their thinking to accept that the electromagnetic field was its own peculiar medium, though generations of college students felt abandoned when they no longer had the Ether to serve as a mental crutch for understanding how light worked. What they now were asked to imagine instead was that fields were their own medium—whatever that was.

The primacy of the field as a component of Nature required the development of a new intuition and a new mental image of the Void.

But the benefits of the "field theory" approach to physics were hard to dismiss. The most spectacular implication was that the older problem of ghostly "action at a distance" went away completely. Action at a distance was the reigning paradox in physics that had remained unresolved for centuries. This is the puzzling mystery of how forces work, mentioned earlier in regard to the magnet and the bowling ball. Although two bodies were not in physical contact, they could still affect one another via a mysterious gravitational or magnetic force. What had now changed was that invisible fields did the work directly. Everything was embedded within some kind of field, whether it was gravitational or electromagnetic. Fields could be created in remote locations in space, spreading their influences at the speed of light to the shores of any object embedded in space. All that was left was the Void, containing the shifting patterns of many different fields, with matter sprinkled like tide pools along the borderline of this ocean. Here's how it would all work:

As soon as an electron flashes into existence, its electric field (try not to think of what this is yet) spreads out in a bubble at the speed of light in all directions like the ripples from a stone dropped into a still pond. In a few seconds, the edge of the expanding spherical field reaches the Moon. In a few years of travel, it reaches the nearest stars. One moment, you are outside the "sphere of influence" of the electron; the next moment, you are inside it and being jiggled by the force it is producing right where you are located. After 100,000 years, the edge of this electric field leaves the Milky Way galaxy altogether. Now suppose you move the electron from one spot to another in space. When the electron is moved, a kink appears in this outflowing electric field, because like the spokes on a bicycle tire, the field is now emanating from a hub located at the new location. So now there are two outgoing fields moving out into space at the speed of light. The older one reflects where the electron was before it moved; the newer one tells distant observers where the electron is now. This would be all there is to the story except that Maxwell says that a changing electric field causes a changing magnetic field to appear. So, the kink in the electric field creates a magnetic field, and now this combined electromagnetic disturbance also travels outward from the electron at the speed of light. It is at that point that we see the kink as an electro-magnetic wave. The Ether is gone, and only a field moving in the Void is left as the medium that transports light and electromagnetic forces from place to place.

Although the idea of the Ether has been discarded during much of the twentieth century, we haven't managed to wean ourselves from the idea of invisible things prowling the Void when we think of gravity or electromagnetism. We have accounted for magnetism as a feature of some invisible electromagnetic field, but we have not actually ex-plained what this field is all about. How, exactly, do these fields work? From what are they made? What happens to a vacuum or to space it-self in their presence or absence? Is there no way we can break this chain of invisible "causes," to actually see fields directly? Perhaps not.

Shortly before his death in 1894, Heinrich Hertz wrote *The Princi-ples of Mechanics Presented in a New Form*. Ten years before Einstein, Hertz also tried to explain Maxwell's electromagnetic theory and its relationship to Newton's physics. What he actually uncovered was a rather thorny problem in physics itself. What was then known about the visible world was not enough to give a completely logical and self-contained description: "If we wish to obtain an image of the uni-verse which shall be well-rounded, complete and conformable to law, we have to presuppose behind the things we see, other invisible things. [We have] to imagine confederates concealed beyond the lim-its of our senses."

What Hertz was driving at was this: When physicists tried to ex-plore how the world worked, they were also forced to propose invisi-ble things such as Maxwell's Ether to provide them with a mental vantage point to describe what was really happening. This is the same vantage point that Descartes and even Newton needed as they thought about gravity and invisible particles in the Void. At the very least, you had to accept that the Void was not empty, even though every single atom could be extracted from it to make it *seem* empty. Some invisible substance or agent was needed just to explain how matter could transmit its electromagnetic and gravitational influ-ences across what would otherwise be empty space. Virtually all of the physical insight about the Void yet to come during the twentieth century would run into this same dilemma. By the time the details had been worked out, Descartes had been vindicated about his clouds of invisible particles, and Hertz's invisible confederates had been discovered lurking in Nature's blind spot. They were tucked deep within the workings of atomic-scale matter. Whatever these things were, they preferred to remain well hidden within the cloak of a new form of logic that defied even our best intuitions. To learn more about them, we would have to dig very deeply into the nature of solidity and understand the way our minds create explanations for things we cannot see.

3

BLIND SPOTS
Quantum Fields and the Physical Vacuum

*I am Shu, the god of unformed matter. My soul is
God. My soul is Eternity.*

—Egyptian Book of the Dead, Sixth Dynasty

On a mild winter's evening in California during 1969, I was sitting in the kitchen with my parents. Papa Sten had just bicycled home from work at the DeLaval Turbine Company in San Leandro. Mama Sten, as always, had made sure her delicious coffee aroma filled the room along with the fragrance of fresh-baked coffee bread. We were chatting about the day's activities and the latest news headlines, when the conversation turned to our relatives in Sweden. Despite the many years since emigrating in 1955, we were always in close contact with family members left behind. What was especially enjoyable was reliving the events from our summer-long vacation with them in 1966, which for me at the age of thirteen had been an eye-opener. We were talking about Uncle Gosta when suddenly the doorbell rang. Our doorbell was not the normal kind that simply chimed once. My parents were fond of musical chimes. This particular chime sounded the eight notes of Big Ben in London. A ringing doorbell at our home was not a rare event for a family with a teenager and two parents active in a variety of community programs. But this time it did cause a very unlikely response. Without a moment's hesitation before "Big

Ben" had sounded its fourth note, I solemnly announced, "Someone has died in Sweden." It was an odd thing to say. We hadn't been talking about death or about obscure Swedish folk traditions at all. I walked the few footsteps down the hall from the kitchen and opened the front door. It couldn't have taken me more than a few seconds. The last note from the chime had just sounded as I opened the door. No one was there. Not a single soul walked the sidewalk. There were no cars on the street. We dismissed this event as some odd prank played by someone, despite the fact that no signs of anyone were ever found. Besides, we had never been the victims of such mischief before. Then a week later, a letter came in the mail from Grandma Sigrid in Sweden. She wrote that my father's best friend, Nils Nilsson, had died the same day the doorbell had tolled for us.

You may have had similar kinds of puzzling experiences. Confronted by an odd collection of events or conditions that should have been explainable somehow, you only come up empty-handed. There seems to be missing information that prevents you from creating an orderly history of events. The reason for this is very simple. No matter how careful you are as an observer, you only have a limited knowledge of the many factors that add their own links to the chain of cause and effect. These links can be hidden at the molecular or the microscopic scale as a bit of invisible corrosion in a switch causing a current to flow through a circuit. They can be located in odd corners of space and in instants of time that you are not able to directly experience at the moment. There are, after all, regions of space within your own home that you have never visited. If someone were hiding in a nearby bush in the dark, you would not know that unless you walked over and inspected the bush. There is another problem that comes into focus as well. We tend to think of ourselves as objective observers of Nature, but we are not omniscient. In fact, we are not even very accurate in seeing the things right in front of our own faces. How can we hope to understand something as exotic as empty space if we can't even be sure our minds are privy to all the information Nature is providing us?

We humans have many kinds of sensory deficiencies that make it hard for us to experience the complete world even when we make a conscious effort to do so. Our brains fashion for us an interpretation of the world that can sometimes be at odds with what actually exists. When no definite explanation suggests itself, we create a growing collection of hypothetical explanations that become less likely as the amount of missing information increases. The retina in our eye, for example, doesn't register everything that falls on it as we look at the scenery around us. At the point were it connects to the optic nerve in the fovea, there are actually no light receptors at all. Your optometrist can show you this dime-sized "blind spot" during your annual checkup. This spot conceals information from us as we look out at the world, and we never know it. Our brain guesses what the missing information might be, filling in what the fovea prevents us from seeing. This is actually a minor loss of information that the brain has no trouble recreating, and we hardly have any sense of the patchwork that has been put in place. Another amazing thing is how shallow our perception of detail can be. If you hold this page right in front of you at arm's length, you can understand the text you are directly seeing. But you cannot make sense of the text if it is located more than an inch away from your direction of gaze. Your peripheral vision can detect the slightest movement, but you will have trouble reading even the largest text font. The retinas of frogs are even more aggressive. They only respond to movement and do not even allow the frog's brain to think about stationary features. Humans can encounter problems with making sense of the world that are even more peculiar than a frog's. Sometimes the human brain guesses incorrectly, with spectacular misunderstandings.

Take the example of phantom limbs, where amputees continue to have direct, conscious experience of arms or legs that no longer exist. These patients are not stupid, nor are they suffering from wishful thinking. As Dr. Ramachandran also points out in his book *Phantoms in the Brain*, "Your own body is a phantom, one that our brain has temporarily constructed purely for convenience." The great variety of phantom limb syndromes, with their origins in well-understood synaptic circuitry, and the many tricks we can play with our senses to confuse what we mean by "self" and "non-self" support such a radical view. We cannot overlook the importance of what the mind is doing to create for us a meaningful world from sensory information.

A retinal blind spot and a missing limb are familiar things that the brain can rebuild from synaptic connections to create a working internal model. But the *meaning* that you assign to your perception of the world and your sense of reality about it are the result of a model created by the mind from moment to moment. Without this model, all the information that floods your senses would merely fall on a blank consciousness—literally, on deaf ears. If we are going to explore the Void, we first have to understand where it is in relationship to *ourselves*.

Without a particular brain structure called the "orientation association area" (OAA) located in the top, rear portion of your head, you cannot distinguish between self and non-self. In their book *Why God Won't Go Away*, researchers Andrew Newberg and Eugene d'Aquili at the University of Pennsylvania have turned up some surprising details about how this brain region works and why it may be at the root of a common religious experience. Apparently, this region, which straddles both the left and right cerebral hemispheres, has neurons that respond only to objects within an arm's length of your body. Other cells respond only to objects located out of reach—the first step in discriminating between things in the universe that are a part of you and things that are not. The left-hemisphere component creates a mental sensation of the physical body based on touch, hearing, and vision. The right-hemisphere portion specializes in locating your body in physical space. When the sensory stimuli to the OAA are reduced, as in meditation, the mind loses its discriminator between self and non-self and the impression is that your body has become "at one" with the universe. So, the sense we have of space, limitless space, and the majesty of the night sky is in some way based upon specific ways that our brain distinguishes self from non-self. Yet even this model-building process has its limitations.

Research conducted by Ramachandran demonstrates how malleable our model-building process can be. Without an as yet to be identified structure in the left hemisphere, you will not be able to synthesize sensory experiences into a coherent internal model of your body's state, or the state of the external world. An analogous feature in the right hemisphere, if damaged, will not allow new information to update the belief system so incessantly and vocally created by the left hemisphere. According to Ramachandran, clinical studies of a condition called "anosognosia" have their origins in such damage. Right hemisphere damage during strokes causes intelligent, rational people to sincerely believe that their paralyzed arm doesn't exist or that it actually belongs to someone else. They will tell you this, they will sincerely mean it, and they will not comprehend any other option. So, given the tendency of our brains to manufacture and confabulate stories about the world to fill in missing details, how do we keep ourselves on an even keel when investigating the invisible? The Void represents an extreme challenge to our senses, and to our mind's ability to construct a model for it, both because of its intangibility and, paradoxically, its ubiquity. The difficulty of surmounting the challenge of filling in the Void's sparse data with a sensible model is compounded by the fact that Nature is itself ambiguous. It is not entirely the fault of our senses or our technology that we cannot form a clear picture. During the twentieth century, physicists learned this lesson well as their measurements and experiments probed the shorelines of the atomic-scale universe. In this microcosm, we have objective physical evidence that the physical world breaks down into another kind of reality. It is a reality that obeys none of the laws of Nature our brains have found so reassuring. We still don't know how to describe this atomic world in intuitive terms because only tedious mathematical relationships seem adequate to the task. It isn't a world we could ever see, even if it were possible to build the most powerful microscope with unlimited magnification. As physicists explored the edges of atomic space where solidity fades into the Void, they discovered that they could no longer see the new landscape with crystal clarity. There were also many new rules and laws to learn, though none were akin to the experiences that physicists had in the laboratory-scale physical world.

Most of us learned about basic atomic structure in grade school, but this knowledge is surprisingly recent. The basic "electron-proton-neutron" model wasn't completed until 1931, when the neutron was finally detected. Aside from becoming aware of its constituents, physicists also gained a much clearer appreciation about the sizes of atoms. Atoms are typically about one hundred millionths of an inch

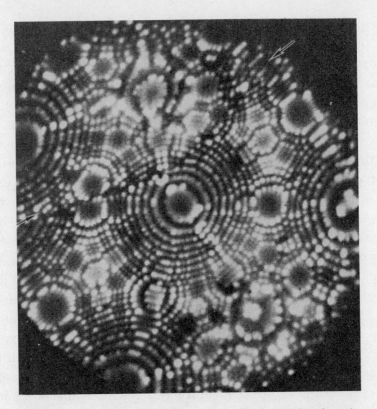

FIGURE 3.1   The world is a mixture of existing and nonexisting elements revealed in intertwined patterns of light and dark. At the atomic scale, space and matter work together to create geometrically regular patterns. This Field Ion Microscope image of a tungsten needle shows individual atoms arrayed in their geometric crystal patterns. Physicists think of this inner world as not really existing in the same way as a table or the book you are reading. (Courtesy of Lawrence Berkeley National Laboratory)

across. If an atom were magnified to the size of a basketball, the nucleus would be smaller than a grain of sand at its center. An atom is really nothing more than empty space with a handful of smaller particles that give it heft. If that is the case, the solidity of a car or a needle like the one shown in Figure 3.1 is really only an illusion, though fortunately with a bit more objective reality than a phantom limb.

The discovery that matter was granularized into atoms and that atoms were themselves made of still smaller components led to a steady stream of discoveries that chipped away at the seemingly clear distinction between the Void and our otherwise solid world. It was especially exciting that the atomic graininess of Nature was also found to be a more universal feature of how things were put together, and not just a feature of solid matter. Plate 3 shows how other systems favor cell-like arrangements. At the same time that atoms were seriously thought of as the building blocks of matter, a different kind of graininess began to appear—this time in the properties of light itself.

When a solid body such as an iron bar is heated, it changes color from red to orange to yellow. It does this every time, in exactly the same sequence of colors. At the same time, it also gets brighter. When you use the burner on an electric stove, you see this for yourself as you turn the dial from the low to the high setting. First it is a dull red, then it brightens to a brilliant orange. When the filament of an incandescent bulb is turned off and cools, it retraces this color shift each time from blinding white to yellow, orange, and finally red. Why does light consistently change in this way as you change the temperature of the hot plate or filament? Many physicists worked on this problem during the last decades of the nineteenth century, but it was the German physicist Max Planck (1858–1947) in 1900 who finally came up with the right explanation. There was, however, a price to be paid for the new insight. Planck's explanation relied on a careful mathematical bookkeeping of the way the electromagnetic field waved at each frequency throughout the electromagnetic spectrum. The bookkeeping mathematics could only work with its dazzling accuracy by switching from quantities that were smooth to ones that were broken down into discrete quantities like the pixels in a digital image. Up until this time, physicists only bothered with describing Nature in terms of smooth quantities that could be indefinitely divisible, such as time, energy, or temperature. That led to heated bodies glowing with infinite brightness. Only by pixelizing light would this mathematical problem vanish. There was something about light that had to be treated as though it came in packets of dotted, pixelized energy called "quanta." This new idea, required by the math, didn't bother Planck at all, because he was more interested in solving a technical problem in mathematics than in worrying too much about its fundamental implications. The idea sure troubled everyone else, though.

This discovery has always amazed me, from the first time I sat in a physics classroom in college and had this mathematical "trick" explained to me. Here you have someone doing nothing more than playing with mathematical symbols on a piece of paper. He tries one approach after another to attempt to bring theoretical calculations in line with simple color-change observations, and nothing seems to work. The math still predicts that light given off by a heated body should become infinitely bright as it is heated, suggesting that light energy is something smooth like water flowing out of a pipe. Planck then switches to math that treats light as something inherently pixelized. This not only eliminates the problem of infinite increases in brightness but also exactly reproduces the way the light spreads out in a finite spectrum of frequencies matching the measured spectrum of bodies heated in the laboratory. Planck didn't make any observation whatsoever to suggest that light was pixelized. Instead, he worked exclusively with the math to uncover this startling fact. I find it just a little bit bewildering and awe-inspiring that by working in the abstract world of mathematics, major discoveries about the physical world can turn up so easily. This could never happen unless the world was firmly anchored in relentlessly logical principles.

Planck's idea was deeply puzzling to physicists at the turn of the twentieth century because, given the work of Huygens and Maxwell, there didn't exist a dispute about the nature of light. It was considered to be a "closed book" in physical science. Light was an electromagnetic wave that rippled through the Ether. It had nothing to do with particles that behaved like Planck's quanta. Even though Planck had invented the light quantum idea, there is evidence in his writings that he didn't really buy into this idea completely. For him, it was not much more than a mathematical "trick." In modern-day terms, it is something like buying a package of hot dogs or a six-pack of soda. There is nothing about either the meat or the soft drink that came in fixed amounts. Both substances could be bartered in cans or links. So how was this contradiction between light as a wave and light as a particle resolved? How could it be that light could appear as Maxwell's electromagnetic waves under some conditions and as Planck's quantum (eventually called the photon) under others? There wasn't much time to think deeply about this problem, because within a few more years, a new discovery overtook physicists. It suddenly became clear that Planck's quantized waves weren't just a property of light alone, but were a feature common among all of the things that could fill the Void.

The Danish physicist Niels Bohr (1885–1962) presented the physics community with an atomic theory in 1913 that seemed to explain most of the available data on the hydrogen atom. There was, however, a catch to this success that went against everything then known about matter. At the time, electrons were thought of as tiny spheres of solid matter that carried their singular electric charge on their surfaces. It was such an intuitively satisfying idea that, like the old Ether idea in its heyday or the artist drawing ears on whales, the accuracy of this mental image seemed certain. What Bohr proposed was that these solid "planetary" electrons could only move in specific orbits within the atom. If they really did look like microscopic planets, why couldn't they orbit an atomic nucleus any way they pleased, just as ordinary matter behaved? This quickly led to a more serious paradox. Despite the fact that the electrons whirled around with millions of times the acceleration of gravity at the Earth's surface, they refused to emit so much as a smidgen of energy, as Maxwell's theory demanded they must. In some unfathomable way, the space inside the atom must be divided into some zones where the electron could exist, and some where it could not. In those "empty" zones, the ordinary laws of Nature were completely suspended. If this didn't happen, all of the matter in the universe would instantly collapse as the supporting electrons were forced to spiral into the nuclei and vanish. But why should the electron only exist in these particular regions of atomic space? Had something strange happened to space itself at the minuscule scale of the atom? Physicists would soon discover that it wasn't space that had changed in such a dramatic way. It was matter itself that had revealed its most bizarre side at these small scales.

One of the most well-known properties of waves can be witnessed with a jump rope attached at one end to a door handle and shaken vigorously at the other end with the right cadence. Like a plucked violin string, you can get many different waves to "stand" on the rope for a split second, making it appear motionless. The easiest one has one full arch between the ends of the rope. The next-hardest standing wave to shake into existence is the one with two arches and a stationary point in between them. It's an amusing childhood game that has probably been played for millennia, but in this common activity there were clues to one of Nature's biggest secrets. The French physicist Prince Louis de Broglie (1892–1987) discovered that if you imagined that matter had wavelike qualities, like the standing waves on a jump rope, Bohr's orbits would have a natural explanation.

Instead of a vibrating rope anchored at its ends, imagine in your mind's eye that with the rope in motion, you pulled the end points of the rope close together and actually tied them. The ensuing waves have to be just the right length so that they fit exactly along the circumference of the rope. If the wavelength were slightly off, the wave running around the string would interfere with itself and fade away into a discordant mess; the standing wave would vanish. What de Broglie was proposing was nothing less than matter behaving in the same way, with these same wavelike tendencies. It was an astonishing idea, and it flew in the face of the traditional thinking of the time. How could an electron that everyone assumed acted like a minuscule BB shot also be spread throughout atomic space? Only two years after de Broglie's ideas appeared in the technical literature, experiments confirmed this extremely unusual idea. Just as light could be diffracted as it passed through a narrow slit, a beam of electrons showed exactly the same pattern of light and dark bands. Electrons could, indeed, act as waves and exist only in certain regions of the atom. The good news was that these seemed to be exactly the regions that Bohr's model required. The bad news was there was a second, odd thing about this new picture of the electron and atom.

Bohr's atomic model treated the electron as though it were a tiny planet orbiting a sun—the atom's nucleus. De Broglie's matter waves as applied to the electron said that this couldn't be so. The electron seemed to be everywhere all at once in a shimmering cloud of indecision, as we see in Figure 3.2. What the electron was doing inside an atom really wasn't motion as we know it at all. It was much more like the wave you see when you pluck a violin string. The physical qualities that make up an electron are present in many locations through-

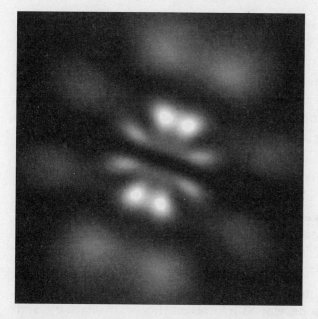

FIGURE 3.2
The wave function of an electron in an atom, showing its wavelike ability to be in many places at once. (Courtesy of Dean Dauger, from *Atom in a Box,* available from Dauger Research, http://dauger-research.com/)

out the atom, though nowhere in particular. Most of the time, the electron exists in the places predicted by Bohr's orbits, but there was also a small chance that it could either find itself inside the atomic nucleus or be a million miles away. This indistinct quality of the electron also meant that the otherwise empty space within an atom is filled by the electron at one time or another, just as a plucked violin string is filled by its own vibration. The space inside an atom is hardly a vacuum in the strictest sense of the word. The recognition that the electron behaved as a wavelike object inside the atom also forced physicists to face another implication about the atom. In many respects, what goes on inside an atom isn't really a part of our human-scale world at all.

Niels Bohr, also one of the founders of quantum mechanics, went as far as to say that electrons and atoms do not exist as "real" things in the same way that trees or tables do. The entire concept of what we mean by "real" or "reality" is up for grabs in the atomic world. According to the new rules that operated in this microcosm, until you made a measurement that registered the position or speed of a particle, neither of these qualities actually existed. Atoms were in some sense pools of potentially real matter embedded in the Void, with no definite shape, structure, or attributes until *after* they were detected. As in Plate 4, even ghost atoms can be created, which are less substantial than atoms themselves. In the face of this irrefutable quantum logic, the very foundation of a durable, solid world familiar to our conscious minds dissolved away into a murkier state of mere possibilities. This didn't mean that you could suddenly stand up and walk through a wall. It meant that the activity inside atoms was ill-defined. In fact, it is so ill-defined that although the insides behave as though they contain objective qualities like mass, spin, and energy, these qualities do not actually exist until *after* you set up an experiment to measure them. As we have explored the tide pool of atomic matter, we see its interior as though through cloudy water, and the pool blends into the Void at its edges.

By the time atomic physicists managed to include matter waves into their new "quantum theory" of the atom in the late 1920s, Einstein had long since developed his special theory of relativity, and physicists were forced to include these new relationships in their descriptions of the atomic world as well. This led to many technical difficulties that were resolved only by completely changing how to think about the Void: what now came to be called the *physical vacuum*. Although their internal laws and structure were peculiar, atoms were still a part of the physical world that obeyed the principles of special relativity, a rubric that provided a detailed prescription for how different observers would experience space, time, and the motion of matter at high speeds. It also prescribed that there would be changes in the measurement of time, space, mass, and energy—all of which were observable quantities central to defining atoms and their constituents. Physicists could elect either to claim that special relativity did not apply to the atomic world (a plausible though uneconomical idea) or try to find some new "relativistic" theory of the atom that would blend the two theories together into a much more powerful and more broadly applicable theory. During the 1930s, several groups of physicists chose the latter option because Einstein and others had found a variety of persuasive objections to the former suggestion. It was through this intense study that the nature of field and vacuum became crystal clear for the first time in centuries. There would be at least one casualty, however, and the most severe one involved one of the most established laws of physics: the conservation of energy.

We all learn in grade-school science classes that energy can neither be created nor destroyed, only transformed from one kind to another as a system changes in time. Although the inner workings of atoms seem to run by some very peculiar laws, it was generally assumed that the conservation of energy still worked. This means that if we calculate the energy of an electron—call it "E"—this should be the same at any time as long as the electron has not been affected by some outside agent like a collision with neighboring particles. The trouble is, in relativity it is not the total energy, E, of a particle that stays the same from moment to moment. Instead, it is the square of its energy: the product E multiplied by E. There is no intuitive reason that the square of the energy of a particle should be a critical, natural quantity. This turns out to be just another one of those perplexing rules in special relativity, like the absolute constancy of the speed of light, that has to be accepted at face value. From basic algebra, E x E (that is, E^2) will always be a positive number, but this can happen in exactly two ways. We can multiply $(-E) \times (-E)$ or we can multiply $(+E) \times (+E)$. In other words, if all you are allowed to know about something is the value for E^2 (say 4.0), then $+E$ and $-E$ will correspond to the positive and negative square roots of E^2 ($E = +2.0$ and $E = -2.0$). But what does this mean physically? A car in motion possesses only positive energy, so this means the negative-energy root can be discarded. For an electron inside an atom, both roots are needed in the mathematics to make them consistent with the logic of relativity. This is exactly where things start to get very bizarre. Since the negative energy states have a lower energy than the positive ones, electrons will want to tumble into the lowest energy state available to them, like a mountain climber falling off a cliff into the valley below. What prevents all of the electrons in the universe from instantly rushing to fill all of the negative energy states that relativity now says are available? Why haven't all forms of matter in the universe vanished into the mysterious never-never land of negative energy states? There must be something that prevents this cataclysm.

The American physicist Paul Dirac (1902–1984) came up with a simple but controversial solution to this energy dilemma. To make this work, you have to imagine every cubic inch of space in the universe filled with uncounted trillions of electrons, each moving with its own speed and direction with different total energies. A quick look around you would also tell you that these electrons could lead to no visible effects. Because no two electrons can be in the same state according to a fundamental law of quantum mechanics, any real electron would have only an energy not previously taken by one of the other "ghost" electrons flitting about in the Void. Like a chessboard with no open squares, you could not add an additional chess piece to the game. Still, a few of these *negative energy states* might be vacant. According to the mathematics, just as black is the opposite of white on a checkerboard, this vacant hole in the vacuum—a Dirac Hole, would look like a twin to the electron, but with a positive charge.

Dirac's bizarre new theory had made a concrete prediction that empty holes in the vacuum might be real physical particles. Did it pan out? A lightweight positively charged particle actually did turn up in photographs of cosmic ray showers. The significance of this discovery was not understood until after Dirac's paper was published a year later. The "positron," as it was later called, was identical to the electron in all respects but one: It had the opposite electric charge, and when you combined the electron and positron, they would both vanish in a burst of light energy.

This idea of antimatter was revolutionary. It was the stuff of science fiction, but the new model for the Void was even more provocative. Empty space was now filled with an invisible but infinite number of Dirac's antiparticles. These particles cannot be detected. They produce no gravity, otherwise their collective gravity would shred planets and stars asunder. They don't behave in any way like the older notions of the Ether. But under certain circumstances, pieces of this vacuum can be made to enter the real world as a positron. It was beginning to seem as though Aristotle and Descartes may actually have had the right idea after all. Nature really *did* seem to "abhor a vacuum," at least one you could try to define as some kind of perfect emptiness. Only now it seemed that physicists were trying to have it both ways. The vacuum was still empty of visible things, but it was flooded to the hilt by invisible particles. What was also troubling was that these particles had no obvious connection to the concept of the field but bore a vague resemblance to the discredited idea of the Ether. The Ether had been thrown out the front door, but now, another band of invisible things was trying to sneak in through the back door.

Dirac's discovery of the positron within his mathematics didn't gain many supporters. Although the evidence arriving daily from the cosmic ray showers was irrefutable, many physicists regarded the positron as an oddity—a particle that just didn't fit in anywhere. The fact that Dirac's prediction involved a bizarre new picture of the vacuum spawned a deep feeling of skepticism within the physics community. After all, it is always possible to predict something new for the wrong reasons. Besides, Dirac's vacuum was a "one-trick pony." It only explained positrons and nothing else. Perhaps more troubling, it didn't have the endorsement of Italian physicist Wolfgang Pauli (1900–1958), the reigning international expert on electrons. Pauli wrote a particularly stern letter to Dirac: "I do not believe in your perception of 'holes' even if the existence of the anti-electron is proved." Could it be that Dirac's prediction was just a lucky shot, so that one had the logical option of dismissing the physical vacuum with its sea of filled holes? That posed a problem all its own. If relativity didn't apply to the atomic world, then someone had the equally difficult task of explaining why this was the case. There was nothing in Einstein's theory that stated it would not apply everywhere—inside an atom or at the very the limits to cosmic space. For a decade or more, physicists remained unconvinced that the physical vacuum looked anything like Dirac's crazy model. Either more experiments had to be found that pointed directly toward Dirac's vacuum, or the mathematics of the vacuum had to be cleaned up so that it could provide more testable predictions. As it would turn out, both of these requirements were satisfied—but only after World War II. After the war ended, physicists not only found themselves back at work in lecture halls across the world, they were also the beneficiaries of a windfall of new technologies that were declassified. In 1947, American physicists Willis Lamb (1913–) and his student Robert Retherford (1912–1981) at Columbia University used some of this technology to make an ultraprecise measurement of how much energy an electron actually has when it is inside a hydrogen atom. This one experiment did more to usher in a revolution in thinking about the Void than even the discovery of the positron. Dirac's original idea of the Void didn't survive this revolution, however. A newer idea about the Void was about to arise from the ashes of the older model.

Strange as it seems, in some areas of science the best way to make new discoveries is by making more refined measurements of those things that are well understood. When an electron resides in an atom and contains the least possible amount of energy, this energy is called its "ground state energy." An electron cannot contain *less* than this amount of energy. The ground state energy of the electron in the hydrogen atom was believed to be a very well-defined number, on par with the mass of an electron or its charge. But this isn't what Lamb and Retherford found when they explored beyond the next decimal point. What they found rocked the physics community. Just when physicists assumed they knew just about everything about the hydrogen atom, Lamb and Retherford announced a new kind of energy shift for the electron in its ground state. It was a minuscule shift in energy, one that was not predicted by the most sophisticated description for atomic physics at that time. It was like discovering a missing lower rung on a painter's ladder. Not even Dirac's theory with its positrons and vacuum holes was able to offer an explanation.

The only explanation that seemed to work came from a fledgling collection of ideas still in search of a formal mathematical description and larger theory. It was such a new way of looking at atomic physics that it barely had a name at all. Some called it "second quantization" or the "theory of the quantum field," but most researchers just took to calling it "quantum field theory." Even in its newness, there were still unmistakable elements of older ideas peeking through, but now they wore a new set of clothes, and like Faraday's lines of force, this made them friendlier to mathematical description. In quantum field theory, fields, as Descartes had proposed, were invisible clouds of particles (now called field quanta) that buzzed about between particles of matter to produce forces. When they were exchanged, they caused pushes and pulls that were felt as forces. For the electromagnetic field, these field quanta were none other than the photons of light that Planck had proposed in 1900.

The electromagnetic force is actually transmitted by the exchange of photons. These photons surround all charged bodies like a swarming hive of bees, constantly being created and destroyed within the electron's field. But these are not ordinary photons like the ones that carry light or radio energy from place to place. You don't see matter surrounded by a luminous aura of light. This is because when these photons pop into existence even for a moment, they totally violate the basic law of physics: the conservation of energy. To avoid a conflict with the conservation of energy, the photons must somehow be undetectable even as they are being created "out of nothing" from the fabric of the electromagnetic field itself. Let's have a look at a specific example and see how this works.

Before the photon appears, the total energy of the pair of electrons is, say, 287 units. When the photon materializes, the energy of the two electrons plus the photon is now 288. This isn't the same amount of energy you started with, so energy has spontaneously appeared out of nothingness and violated the energy conservation law. These "virtual photons" can use their temporary existence to transmit invisibly the electromagnetic influence of the electron over a great distance. Upon absorption by another electron or charged body somewhere else in space, the virtual photon vanishes and imparts a kick, which is felt as the electromagnetic force. But if a virtual photon could receive enough energy to promote it into a particle with positive energy, it would enter the world as a real photon of light, a wave in the electromagnetic field. It's very easy to provide this missing energy. All you need to do is turn on a flashlight, or if you happen to be a physicist, slam two atoms together. Where did this stolen energy come from? It came out of the vacuum itself.

The German physicist Werner Heisenberg (1901–1976) discovered a way in which small "embezzlements" of energy are actually built into the strange world of the quantum universe—as long as they don't go on for too long. The reason this happens has to do with the hazy way particles move in space and time. They don't move like BB shots, but like waves of energy spread out in space and time. This creates a built-in limit to how well you can know both a particle's speed and position. When you look at the standing wave on a jump rope, the location of the wave is spread out along the rope and you cannot say it is in any one specific part of the rope fiber. This means there is also a limit to how well you can ever know the total energy of a particle or system at a specific instant in time, and that is the basis for the law of conservation of energy. The conservation law only works in an average way. If you look at a system too closely, the energy transactions don't always balance from moment to moment. Fortunately, Nature prevents us from ever seeing these transactions in progress, and this is why the fields that are responsible for forces such as magnetism are also invisible. An analogous kind of uncertainty can also be found in very large systems.

Consider the inherent uncertainties found in another process that is supposed to have a clear outcome: national elections. Every voter casts a unique vote. The expectation shared by everyone is that by tallying these votes, a unique presidential outcome will result. In the U.S. presidential election held in November 2000, votes were entered into the machines to be tallied, and several hours later, a decision came out the other end. Then, the way in which the votes had been tallied was examined in detail to find the exact logical pathway taken in a single state: Florida. It was at this point that the apparently clean outcome of the election was clouded by new knowledge of just how complex the voting process can be. This information wasn't of interest in previous elections when the vote counts had ample margins to establish the outcome. An initially clean measurement grew increasingly more ambiguous the more that the process was scrutinized and the more criteria were applied to counting individual votes, until thirty-seven days later, when the U.S. Supreme Court, in essence, elected the president.

Just as the vote tally changes often involved uncertainties of hundreds of thousands of votes, the energy transactions that are hidden in the Void can involve impressive sums of energy, at least for a fleeting moment. Thanks to Einstein and his famous formula $E = mc^2$, we know that mass (m) and energy (E) are equivalent physical concepts, so a big enough uncertainty in energy could temporarily translate into the sudden appearance and disappearance into the Void of an entire particle. Because the vacuum has to have no net mass or charge, Dirac's antiparticles are needed to logically round out the picture. This would provide the circumstance that if enough energy uncertainty existed, a particle and its antiparticle could appear out of the Void and travel through space for a short distance before they would collide and annihilate into a blip of pure energy. The journey would be a short one if the energy was just sufficient to create the pair and no more. An electron and a positron can live only a scant fraction of a second, and even at the speed of light, they travel far less than the diameter of an atom. In our basketball-size model of the atom, their journey would be barely 3 millimeters—the size of a rice grain. Fortunately, Nature won't allow us to observe this violation of energy conservation, no matter how cleverly we set up the measuring apparatus. Because they exist in a much hazier state of reality than we humans, electrons respond to these virtual particles as they appear and disappear in the Void. This causes slight deviations in their energy, which is why the Lamb Shift occurs. Nature does not completely prevent us from observing the electron, nor from seeing in its mirror a glimpse of virtual particles. It just won't let us see the transactions *directly*.

Despite the apparent successes of this new way of thinking, for a short while, few physicists really subscribed to it. Quantum field theory seemed like an explanation that accounted for the Lamb Shift and nothing else. As Nobel laureate in physics Steven Weinberg noted, "There is a huge apparent distance between the equations that theorists play with at their desks, and the practical reality of atomic spectra and collision processes . . . The great thing accomplished by the discovery of the Lamb Shift was not so much that it forced us to change our physical theories, as it forced us to take them seriously." It took a decade before physicists finally put aside their skepticism about the new view of the vacuum and how particles could be affected by it—a much shorter time than it took to discard the Ether. For physicists, nothing succeeds like success. The results of decisive experiments usually persuade even the most stubborn minds. That's why the detection of the Lamb Shift was so pivotal and persuasive. It showed that something odd really was going on, and there already was a theory available to explain the new discovery. If the new theory had only predicted the Lamb Shift to an accuracy of one or two decimal places, many would have dismissed it as a lucky guess. What do you do when a bizarre theory predicts and experiment verifies Lamb Shift energy to ten decimal places? Once a set of technical issues was resolved in the early 1950s, there was no longer any opposition to the mathematical aspects of the new theory. Quantum field theory grew in stature to become *the* theory for describing how charged particles affect each other. Its electromagnetic version was called quantum electrodynamics or, simply, QED. Other versions were soon deployed to begin the harder task of understanding the other two atomic forces: the strong nuclear force and the weak nuclear force, the two forces that hold atomic nuclei together and cause some particles to decay. For the first time since Descartes proposed the idea over 300 years ago, a major logjam had broken and the pace of discovery accelerated. Physicists still couldn't see directly the virtual contents of the Void, but it almost didn't matter because they could use their mathematics to anticipate all of the ways these hidden constituents could animate matter. Like the way the brain fills in a blind spot with detail or "creates" an amputated limb with a mental phantom, physicists could mathematically work around the missing information denied by Nature and still see a complete, logically consistent world beyond. American physicist Richard Feynman (1918–1988) even developed eye-catching diagrams for showing how a wide range of electromagnetic phenomena work in terms of virtual particles operating in the invisible Void.

A delightful feature of these diagrams, like the impressionistic rendition in Figure 3.3, is that they seem to provide an accurate method of drawing a picture of the contents of the physical vacuum. Feynman diagrams have a curious beauty to them. They show electrons drawn as solid lines with arrows pointing forward in time to represent the history of the electron as it moves through space. Photons, by contrast, are drawn as wavy lines. The crossing points, like knots in a spider web, tell about the instant when the photon (electromagnetic field) interacts with the particle. In fact, Feynman diagrams, as literal as they were, were still nothing more than symbols for various terms

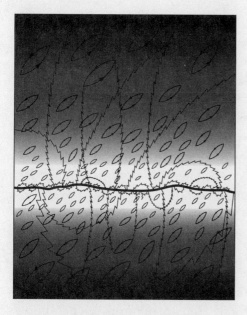

FIGURE 3.3
The physical vacuum revealed by quantum mechanics contains virtual particles that come and go, affecting the matter we see by imparting their many pushes and pulls. None of this activity can ever be directly observed. Empty space is a quantum network of things that come and go under the cloak of perfect invisibility. Physicists have detected this hidden world by carefully studying how elementary particles like electrons behave as they occupy space and move through it.

in an equation. They could just as easily be thought of in less provocative ways than the things they seemed to represent. In electronics, for example, specific but arbitrary symbols are likewise used to represent resistors, capacitors, and transistors. The symbols only vaguely have anything to do with the activity of electrons as they flow around a circuit in a radio or a computer. Yet these symbolic visual patterns help us organize our calculations and visualize the design of devices of staggering complexity involving millions of discrete components. In QED, similar symbolic diagrams help physicists keep track of the thousands of terms and millions of factors necessary to get a testable prediction, and *that* is the bottom line in any scientific theory. In some sense, Feynman diagrams are nothing more than scratch-paper figures drawn as an aid to calculation, much as Maxwell's Ether had been. The diagrams are discarded when the calculation is completed and a prediction is computed. As peculiar as the new Void seemed to be, the new picture that emerged from the mathematics does work. And it does so in ways that simply boggle the mind.

The success of QED also leaves us with a very deep issue to ponder as we try to understand our new relationship to the Void and the physical vacuum. Like a mysterious bump in the night or the doorbell that rang of its own accord, the actual causes of some things in Nature may, in fact, be forever beyond direct observation. That, in itself, does not mitigate their profound impact. What was even more surprising about this revolution in thinking about the Void was that it demanded that physicists accept the existence of invisible components to the physical world even though the physicists' own senses and technology would never be able to detect them *directly*.

In some ways, this dilemma was similar to what some stroke patients endure when they learn to renegotiate their world. Individuals afflicted with anosognosia, though completely rational and intelligent, are consciously unable to "think their way back" to a rational world where they admit that a paralyzed limb still belongs to them. Unlike the malleable reality experienced by the individual, science behaves like an externalized human brain—shared by a community of individuals in which individual deviations count less than the consensus. In an amazing act of self-transformation, science bootstrapped a new reality out of itself, a reality that now contains unobservable phenomena hidden in the Void. So why don't physicists talk about this amazing virtual world in more detail? Quite literally, they are at a loss for words. In many ways, there just isn't much to be said about this world beyond what the mathematical symbols and Feynman's diagrams provide. This omission has always seemed to me a story worth taking the trouble to tell, or at least worth making the effort. When I have repeatedly gone in search of an intuitive explanation for this phantom world, more often than not I have discovered it is described in rather stereotypical prose.

Regardless of what book you read on the subject, you will always encounter a similar description of virtual particles. Most descriptions begin with a statement that "we can never say that a vacuum is really empty because of Heisenberg's uncertainty principle." Following that, there will be a short discussion about what Heisenberg's uncertainty principle is, followed by a statement that "the vacuum is therefore a sea of continuous, and unobservable, activity." The discussion may conclude with a statement that "we consider virtual particles to be real in the sense that they have measurable effects on the energy levels of hydrogen atoms." There are practically no discussions, either in the technical or popular scientific literature, that say more about virtual particles.

Despite our tremendous understanding of how they operate, we still have no better words to describe what virtual particles are or what they do. We don't even know how to redefine our concept of what is real in order to accommodate them. While they are less real than the electrons we can see and with which they interact, they are also more real than the dreams that animate our internal mental worlds or the phantom limbs that confound the altered mental model of some humans. All we seem to have to describe them is our mathematics. Is that enough to make them real?

In his trial before the Inquisition, Galileo was forced to admit that the Copernican doctrine, with its sun-centered orbits, was simply a mathematical tool to simplify the task of calculating planetary positions in a geocentric universe. As the Inquisition insisted, it did not represent the true workings of Nature. But in the 500 years since then, scientific progress has been inexorable. Mathematics, when it works, is viewed as an authentic mirror of the real world, not simply an irrelevant tool or caricature of Nature. It embodies the essential logical relationships within what appears to be a rational world governed by cause-and-effect rules of logic—the kind of model our brains like to fashion. Mathematics lets us predict new things in Nature we haven't seen yet. This reinforces the scientific conviction that Nature is inherently based on logical rules and the physical manifestations of these rules in the patterns we see. Virtual particles are trapped in this same logic of necessity, and we are left much as Galileo was before the Inquisition.

Since the 1950s, we have learned that there is every reason to accept that virtual particles are bona fide members of the physical world, though of a different order of reality than the chair you are sitting on, or even the ghostly electron inside the bowels of an atom. The mind still reels in its effort to find a way to grasp what they are, but this is no longer seen as a barrier to accepting the implications of the evidence. We know now that there are many phenomena in Nature that defy human intuition. In a manner of speaking, fields of force are a new state of ephemeral matter consisting of individually unobservable virtual particles. Every time you play with a magnet to move metal filings, you are invoking the effects of an invisible physical substance that consists of virtual photons. These unobservable particles are the intermediaries that Descartes said filled the Void and allow causes and effects to happen without any mechanical or direct contact. Heinrich Hertz's vision triumphed. Some fifty years before experiments had confirmed QED, Hertz had already guessed that we would someday have to accept the existence of what he called the "invisible confederates" that exist beyond our senses. They would have to be added to our theories so that, in the end, we could obtain a complete, consistent understanding of Nature. Today, we know that under certain controllable conditions, these invisible confederates can be brought into the light. Descartes and Newton would have felt perfectly at home with this idea. Even Aristotle might have smiled at our discovery of what he had announced millennia ago: "Nature abhors a vacuum." Three centuries after Descartes, physicists had managed to make it a working theory. I find it strange that Descartes came up with an idea that captured the gist of quantum field theory long before the first experiments could hint at its correctness. How could our brains have anticipated such an analogy for an invisible process? What is even *more* striking is that, in order to provide the "proof" of Descartes's assertion, we have had to do nothing less than completely transform our entire civilization to accommodate electricity, and all the other spectacular technologies that have emerged from our mastery of the quantum world. We *are* the undisputed masters of electromagnetism in all of its guises, from radio communication to nano-technology and continent-girding electrical-power networks. Will the "proofs" of other ideas in physics and the investigation of the Void lead to still more spectacular revolutions?

We began this story by wandering among Nature's tide pools and exploring the particular one we call "matter." As the Void lapped up against the edges of this pool, we have seen matter become an indistinct wave. The scintillating electromagnetic field shakes with uncertainty as its fabric is reduced to particles of light and virtual matter flashing in and out of existence. Even without a material source for the field, the quantum twinkling of energy trapped in the vacuum still remains as a sea of untapped energy. We have nearly exhausted the "matter" pool of its clues to the nature of space and have found answers to the questions "What is a field?" and "How does it work?" We have not, however, mentioned gravity and space directly. Space has served only as a container, and gravitational forces have not diverted a single electron from its atom-sized trajectory. Yet the geometric properties of space *do* control atoms and their contents by forcing them to be three-dimensional. We have not encountered an explanation for how matter is related to the vacuum itself, or how it is that matter generates a quantum field. Could there exist a deeper level of detail to the Void than we see in the quantum shimmering of the electromagnetic field? And there are still two other forces in Nature, which may yet offer us some new insights beyond what we have currently found among electromagnetism and gravity: the strong and weak nuclear forces. It is to these forces that we now look to see if they hold more clues to the nature of the Void.

4

PATTERNS IN THE VOID
Vacuum Energy and Hidden Fields

'Twas brillig, and the slithy toves
Did gyre and gimble in the wabe;
All mimsy were the borogoves,
And the mome raths outgrabe.

—Lewis Carroll, "Jabberwocky,"
Through the Looking-Glass

Boston is a town known for its changeable weather. It is a coastal city, a collision point for warm Gulf currents sweeping up from the south and cold air masses passing through from the northwest. On this particular summer's afternoon in 1980, my fiancée, Susan, and I were driving into Harvard Square to do a bit of shopping. The weather had been perfectly calm as we stepped into our car, but within moments, a briskly moving cold front swept in from the west. A torrential wall of water pounded the car and sent pedestrians scrambling for cover from the painful hail of raindrops. Dark clouds rushed across the sky to the east, leaving a clear, sunny afternoon sky in their wake. As we reached the crest of the Harvard Bridge, catching our first unobstructed view of the skyline of Boston, we gasped in wonderment at what we were seeing. The sky over Boston was a dark angry mass of receding clouds reaching halfway to the zenith. Etched like a megawatt searchlight across this scene was a dazzling double rainbow

59

in magnificent colors. Susan and I were soon joined by many cars that pulled to a stop on Harvard Bridge to admire the rare spectacle. Of course, none of us had a camera.

Sometimes random natural events can create moments of sheer beauty, symmetry, and color that seem to come out of nowhere as if they were woven into some unseen fabric. The turbulent dark clouds seem like the last place anyone would look to find a perfect semicircle of light sorted out by its color in a regular pattern. Isn't it a bit strange that no matter where a rainbow turns up, it always looks exactly the same?—a sweeping arc of light, banded by the same patterns of colors from red to blue. Where is this pattern written into thin air? How does the sky over Boston know to offer up such a beautiful geometric shape with the same lovely fidelity as it does over San Francisco or Cairo? Scientists will tell you that it all has to do with the way light interacts with matter and Nature's economical use of a few basic rules. Water droplets suspended in the air have nearly the same physical properties no matter where you are on the surface of the Earth. Light rays are refracted and dispersed by water in much the same way, given similar arrangements of where the observer is standing and where the Sun is located in the sky. At least that's what physics seems to demand. But a rainbow is no more a strictly physical phenomenon than a dream is simply a random firing of neurons. No two people really see the same rainbow because no two people are passive observers of Nature's showmanship. Just after the light rays are refracted by the raindrop but just before we turn our attention to other matters, our brains dissect the scenery and return to our consciousness the complete experience of a rainbow. It is an experience defined as much by who we are as by what we are seeing.

When you study the brain in detail and examine where it receives most of its information, a most remarkable realization begins to emerge. The vast majority of the neurons and synaptic connections have little to do with the outside world or information gleaned from the human senses. The brain spends a lot of time talking to itself, not paying attention to its surroundings in the universe. It is an undeniable feature of the way the brain operates that it sifts patterns out of the cacophony of the senses and uses these patterns to find rules that help to insure its survival. The most dramatic pattern is the time and space filter, which contrasts how the left and right hemispheres of the brain function. This leaves its mark in many of the physical conundrums we encounter: matter versus energy, particle versus wave, matter versus antimatter, and even cause and effect versus holism.

Brain researchers have known since the pioneering studies by Nobel laureate Roger Speers at Caltech that our two cerebral hemispheres extract different levels of meaning from the sensory stream. The left hemisphere, which hosts our language centers and loves to talk, is tuned to finding patterns ordered in time such as sequential logic, counting, and cause and effect. The right hemisphere absorbs the entire experience all at once and immediately codes it with any of a variety of emotional flavors. The right hemisphere is good at seeing the big picture rather than trying to deconstruct it into its components the way that the left hemisphere prefers. Analytical thinking and language are left-hemisphere skills, and it learns by narrow example and trial and error, as well as by following rules. The right hemisphere has trouble following rules and favors rich patterns that have to be understood in their entirety. It also responds to nouns but not verbs and extracts meaning from music and facial features. These traits, when coupled with memory, provide an incredibly rich groundwork of archetypal patterns, motifs, and relationships in space and time that can be used to extract all kinds of meaning from the sensory information the brain receives.

Physicists often talk about "string" or "glue" when they try to explain new areas of the physical world. Given the structure and function of the human brain, they can't help applying patterns found in the everyday world to a completely different arena. Physicists use imagery to explain the most obscure workings of the universe: a shoestring and the mathematical structure of a field blend together in a seamless fusion of relationships. Such imagery is a result of the brain using similar networks of neurons to extract meaning from sensory impressions. Every experience, whether it is an internal intellectual one or an external sensory stimulus, is the start of the next pattern of synaptic firings. There are no chemical tags that differentiate between a true sensory stimulus and one that we create in our intellectual thoughts. And this leads us to the most profound recognition of all. Why are there so few patterns needed in Nature to encompass the kinds of phenomena we see there? Given the unlimited resources of the human mind to extract an almost unimaginable variety of relationships in time and space, why is it that Nature is itself so dismally uninventive? The more scientists have probed the innermost workings of matter and force, the more Nature's frugality has become obvious, though in the delicacy of a rainbow on a summer's day, such frugality can still result in dazzling displays of beauty.

My first-grade teacher, Miss Encke, once complained to my mother that she had given up trying to teach me that human faces had to be tan or brown, not purple or green. I still have the offending artwork in my collection of childhood memorabilia, with its stern, handwritten note from my teacher stapled to it. When my mother asked me afterward why I didn't use some other color, I told her that my crayon box only had green and purple crayons in it. I could create a few believable, natural things like purple eggplants (had I known what they were) or green leaves, but everything else was hopeless. In many ways, Nature works the same way with its limited set of crayons, using similar patterns over and over again to fashion stars, planets, and even atoms. Out of all the fantastic things we can imagine to exist, from fairies dancing on the heather to demons lurking in the dark, Nature cannot create any from the limited crayons it has at its disposal. This is perhaps one of the hardest lessons for the can-do optimist to accept.

The nineteenth century was a blissful childhood for science, when Nature seemed to need only gravity and electromagnetism to account for practically everything that cluttered the physical world. Then, during the twentieth century, physicists uncovered two additional natural forces operating alongside gravity and electromagnetism. These were not at all the kinds of forces you could ever recognize through a casual inspection of human-scale natural events. Yet without them, the world would not exist, rainbows would not grace the sky, and stars would not shine in the heavens.

Deep within the uranium atom, 92 protons and 146 neutrons jostle each other in a frantic nuclear dance. The space is very cramped, barely one hundred trillionths of an inch across. Traveling at nearly one-third the speed of light, the particles go nowhere fast. Thanks to our grade-school teachers, we know that positive charges repel each other, so how in the world could 92 positively charged protons hang together in such a cramped space without blasting the nucleus to smithereens? There must be a force buried within the atom that overwhelms the proton's electromagnetic repulsion, keeping the whole bag of nuclear particles intact. Had it been discovered in the eighteenth or nineteenth centuries, it might have been given some picturesque name. In the no-nonsense climate of twentieth-century physics, this new nuclear force was called simply the "strong force." It is a name whose simplicity and directness seems unnecessarily harsh and utilitarian, especially when you consider the linchpin role it plays in lighting our Sun's thermonuclear furnace or in terrorizing the political landscape tucked as it is into fission and fusion bombs.

Despite the fact that it is the most powerful force in the universe, the strong force has its limits. Nature has a tough time creating elements with more than about 100 protons before anything like the strong force can no longer hold them together. Because protons act as waves just as electrons do, they are slippery enough that the strong force can't confine them to a definite volume of space with 100 percent certainty. The nuclei of some atoms act like leaky bags, allowing protons and neutrons to escape in bursts of nuclear disintegration. Radioactive uranium, for example, gives off a helium nucleus and leaves behind a nucleus of thorium. This kind of decay, though it takes about 4 billion years, is rather simple to describe, given what we already know about the quantum laws that operate at the atomic scale. All you need is simple subtraction to see how the 92 protons and 146 neutrons of a uranium nucleus become the 90 protons and 144 neutrons in a thorium nucleus by ejecting 2 protons and 2 neutrons: a helium nucleus. Nuclear disintegration, however, isn't just a feature of obese nuclei. It can also happen to scraps of matter we previously thought were the most elementary, and therefore immutable, forms of matter.

When left alone, some individual particles—the neutron, for example—decay into simpler and more stable fragments. In less time than it takes to bake a potato, a neutron dissolves into a proton, a neutrino, and an anti-electron, all of which fly off into space. How could a fundamental particle like the neutron just fall apart on its own? There must be yet *another* force in nature that makes the decay of individual particles possible. The weak nuclear force, as it came to be called, is about 10 million times weaker than electromagnetism. What is especially puzzling about this force is that there is no obvious rhyme or reason to explain why it acts on a neutron after ten minutes but leaves a proton intact for more than a trillion trillion years. Although microscopic in their dominion, the strong and weak forces confronted physicists with as many hard questions as their much larger cousins, gravity and electromagnetism, ever had. Meanwhile, although the list of elementary forces now stood firmly at four, the list of new particles of matter seemed to be expanding without any obvious limit. Nature had decided to be frugal in using only four forces, each its own unique paintbrush, but when it came to matter and colors with which it would paint the universe, it had elected to be decidedly uneconomical.

Physicists tried to make sense of the dozens of new particles they had created in their labs during the 1950s and 1960s. Again, they seemed to settle on only one solution to Nature's particle proclivity. The quest for simplicity eventually led them to the idea that these nuclear particles were themselves composed of a still more elementary set of ingredients called "quarks." You never saw quarks individually because the strong force, which is produced by particles called gluons, always held the quarks together inside heavier particles, permanently imprisoning them within the nuclei of atoms. The overriding benefit gained from the long and expensive study of smaller bits of matter was that the many spectacular ingredients of the physical world could at last be tracked down to the interactions among a small set of elementary particles—quarks and electrons— and four basic forces. Physicists had at last reached a Nirvana-like state of simplicity that existed as the bedrock beneath the bewildering complexity of matter and forces. A rainbow on a rainy day and the stars in the sky really were connected together in a subtle way that only heightened their beauty.

With new forces identified and a simpler scheme for organizing fundamental particles in hand, the focus of research turned to a new and still more challenging task. Were *all* the brushes used by nature really different, or could you get by with only one of them used in four different ways? In the beginning, there were attempts to unify electromagnetism and gravity, but this led only to a frustrating fall down a rabbit's hole of abstract mathematics, tempered with very little data. Meanwhile, physicists working in their labs inundated themselves with more data about the strong and weak forces. The time was ripe for putting all the pieces together and finding a unified explanation for how everything worked.

The first step in seeing beyond the diversity of the forces and particles to a common kernel of unifying properties is to construct a similar language. Of the 2,000-plus human languages, all fall into a handful of different groups that share common elements of grammar, syntax, and even etymology. Languages have been used, like DNA, to trace the ancient migrations of humans as their various words and grammars became integrated into newer, more modern languages. But quarks and gluons are not nouns and verbs. If you want to understand how the strong and weak forces work and find their commonality with electromagnetism, how do you even begin such a research program? You do what humans have done for millennia. You take an explanation that worked well in one arena, and you see if it can be made to work in another. You also hope that in doing so, you haven't selected the wrong kind of pattern to impose. As we discussed in Chapter 2, fifteenth-century Dutch painters had trouble painting beached whales because their archetype for animal heads included obvious ears, which are lacking in real whales. Physicists didn't have too far to go to find such a robust template. The logic and conceptual imagery that had worked so well to create quantum electrodynamics (QED) was recycled and applied to explain the weak force.

To make the successful QED pattern work for the weak force, you would have to describe the weak force in terms of Descartes's invisible clouds of particles, and their modern restatement in terms of quantum field theory. The very simplest assumption would be that like the electromagnetic force, the weak force would also be caused by the exchange of invisible particles. Because of the limited range of the weak force, the quantum rules said that these particles would have to be very massive so that their ranges would be smaller than an atomic nucleus. A quick calculation placed their masses at nearly ninety times that of a single proton, about the same as the mass of a single atom of rubidium. This was well beyond the capability of any laboratory to create. Unlike the single photon, which did all the dirty work for the electromagnetic force, there would have to be three particles to account for the specific ways that the weak force caused both charged and uncharged particles to decay: One particle would carry no charge, but the other two would carry a positive and a negative charge.

The most complete theory to unify electromagnetic and weak forces was actually a synthesis of the efforts of three physicists who were among the leaders of this research: Steven Weinberg and Sheldon Glashow at Harvard University and Abdus Salam at the Imperial College in London. Like hikers traversing a wilderness along the same trails, they had independently crafted nearly the same theory in the late 1960s. The basic idea behind the "electroweak theory" was that at temperatures above 1,000 trillion degrees or so, the electromagnetic and weak nuclear forces would blend together to become indistinguishable. This was something that no one had ever seen before, but if you were going to unify two very different forces, something like this actual physical unity simply would have to be the result. If this new theory were proved correct, we would have to accept the fact that forces are a lot more mutable than anyone had ever imagined.

As more powerful "atom smashers" were built and brought to bear on testing this theory, experimenters finally did get to see the electromagnetic and weak forces change their strengths the way the theory predicted. They also discovered in the 1980s the three very massive particles that carried the weak force, which had by then been named the W^+, W^-, and Z^0 particles. What was even more surprising was that these new particles made their entry into the physical world at nearly the exact masses predicted by electroweak theory some fifteen years earlier. Could it be that the phenomenal successes of QED were now being used and extended to cover the weak force, too? It was almost too good to believe. From the standpoint of being able to carry out exotic calculations, electroweak theory had indeed struck a very rich vein of ore. Electroweak theory promised and delivered a dramatic series of successes, but like the Trojan horse, an aspect to the theory that was especially troubling was lurking deep inside the assumptions of the theory. In many ways it made the QED vacuum filled with its invisible particles seem rather tame. Like Maxwell unifying electromagnetism and discovering the origin of light, the new theory unified the electromagnetic and weak forces. As a by-product, it accounted for one of the most common and mysterious properties of the physical world: mass. To do that, it had to predict the existence of a new field in nature that stood toe-to-toe with the ones that carried the four elementary forces of nature. Something called the Higgs field would serve as the catalyst for causing particles to have the quality we call mass and to cleave apart the electromagnetic and weak forces.

Mass is rather simple to think about when you consider how we experience it. It is simply a measure of how hard it is to push or pull something to get it to start moving. Forgetting for a moment the effects of friction, if you have to push something very hard to get it moving or to change its direction of motion, we say that the object has a lot of mass. Despite what your bathroom scale might show each morning, mass is actually a rather minor component of your makeup. In many situations, it seems as though Nature could have done without it. If you were to do a nose count of the kinds of particles in the universe, mass-free photons and neutrinos would dominate over everything else by far. It is amusing to search for the origin of your own body's mass and see exactly where that trail ends. The neutrons and protons, which give the atoms in your body their heft, dissolve away into triplets of quarks. From real-world experience, you can slice a pizza into three equal pieces. Each of these is supposed to have one-third the mass of the original pizza. One would expect that in dividing a proton into its three quarks, the quarks would also share equally the mass of the original proton. They don't. Although protons and neutrons have about the same mass, the three quarks out of which each is constructed carry only about 3 percent of this total.

The rest is hidden out of sight in a bank that was set up by Einstein. In special relativity, on T-shirts, and on advertising logos, we come into contact with Einstein's iconic equation, $E = mc^2$, which says that energy (E) and mass (m) are physically equivalent qualities. Inside each proton, the three quarks carry a smidgen of their own mass equal to three percent of the total proton mass, but the rest of what we call the proton's mass is actually in a form of tension-energy among the quarks. The energy carried by the gluon fields is what binds the quarks together inside each proton and neutron. The question isn't "Why do things have mass?" The question is, rather, "Why does nature need this three percent residual at all?" If most of the mass of a proton or a human is contained within the gluon fields that hold the quarks together, why isn't *all* of the mass there in the first place? Electroweak theory blames the Higgs field for this little oversight in Nature. The Higgs field ensures that electrons and quarks are not completely without mass, like their cousins the photons. Were that not the case, our constituent particles would be zipping around at the speed of light. It's that mysterious three percent that literally anchors us to the here and now. Electroweak theory says that the existence of mass is linked to how particles interact with the Higgs field, and like the geology of the surface of Mars shown in Figure 4.1, this field may not be the same everywhere in space. Because the Higgs field causes the weak and electromagnetic forces to become distinct at temperatures lower than 1,000 trillion degrees, this leads to still another strange phenomenon: Particles can actually have variable mass.

Suppose you took your fifteen-pound turkey out of the freezer and put it in the oven. As the oven warmed and the turkey baked, its weight plummeted to two pounds. Taking it out of the oven, it cooled and regained its lost mass. You would think this turn of events quite strange and unnatural, but in the world of subatomic physics, it is just a by-product of how the electroweak theory would actually work if we could observe its action in our kitchens. The theory says that the masses of particles are nearly this pliable, depending on the temperature where they are measured. Electroweak theory shows us a world that has to be tuned like the thermostat on a kitchen oven before it can come anywhere close to looking like the world in which we live.

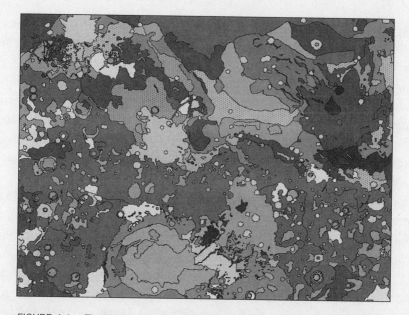

FIGURE 4.1   The Void may be a patina of subtly different domains. These domains may be larger than our own visible universe, and in them the very laws of physics may be different. This map of a portion of the Martian surface reveals its many geological surface types. (Courtesy of USGS, http://webgis.wr.usgs.gov/downloads/mars/geology/m1802b/m1802b.jpg)

Through the equations that define this theory, we catch glimpses of a perfectly unified (and unimaginably hot!) world. Here, electromagnetic and weak forces have the same strength. All the quarks, along with the electrons and their cousins, have no mass at all. In fact, the quality we call mass doesn't even exist in this kind of a universe. It is a classic devil's bargain. Without this bizarre new feature, we logically have no credible way to calculate the W^+, W^-, and Z^0 particle masses that the theory seems to have predicted correctly from among an infinite list of alternate values it could have given us instead.

Adding one more field to Nature and hiding it in the Void doesn't seem like such a high price to pay for a theory that promises and delivers so much. But hiding the Higgs field inside the Void still leaves its weirdness uncomfortably close to our everyday world. It is a field that lurks in your kitchen, as well as in the black vista of space you see in the night sky. Once again, Hertz's warning that we have to make peace with invisible confederates rings true. In every theory of nature that has worked well, we have had to propose some new ingredient to the Void that lets the logic in the theory flow. Without the logic, we cannot make the predictions that match up so wonderfully with what we later uncover in the nuclear world. To deny the logical necessity of the Higgs field is to forfeit the basis for predicting the three new particles of the weak force, their masses, and a slew of other quantitative results of the theory. One might just as well use a dartboard to do physics.

Unlike the idea of the Ether, which never showed up in any of the equations that explained electromagnetism, the Higgs fields appear among the mathematical symbols in a way that forces us to adopt a specific interpretation. Without this specific interpretation, the entire logic of the theory falls apart, no matter how many correct predictions it seems to make. The Higgs field must be a universal field in Nature. It must have exactly the same physical properties in every cubic inch of space and at every instant in time. There is no logical choice in this matter. What is also very bizarre about this field is that there is only one place to put it in the physical world. It must be in the ubiquitous vacuum itself, not in some identifiable source (quark or electron, for example) as for the other fields in Nature. Working with a sourceless field is a nightmare for human intuition. The field is present everywhere, but it doesn't need something like a quark, a flashlight, or a star to produce it. It simply exists. Like Maxwell's Ether, however, it does not impede the motion of quarks or planets through space, and it does not produce gravity.

If the gravitational field were thought of as an angel food cake, the Higgs field would be a thin layer of frosting covering the cake everywhere. Physical particles would be like the raisins and candy plopped into the icing here and there, always in contact with the Higgs field and deriving their masses from it. Like the frosting that can be swirled to create a wedding or birthday motif, the Higgs field also changes the way the Void looks by giving it a distinct patina. The Higgs field acts like the texture of a roughened roadway, or the viscosity of molasses, to force particles to behave in a specific way. Its interconnections are not just limited to real particles. Because the Higgs field affects the masses of real particles, it also affects the masses of the virtual particles that emerge from the vacuum and are the cornerstone of QED. Incredibly, we have a new ingredient to the invisible world that affects both visible and invisible particles embedded in the Void. It's hard enough to think of invisible virtual particles weaving complex patterns in the Void. It absolutely confounds our intuition to grasp how it is that the invisible Void can have a texture. As for many concepts in physics that we have to find words to describe, much of our confusion hinges upon a matter of definition. We expected a whale to have ears because all known heads of large animals have these structures. As it now turns out, Nature has forced us to redefine in our minds what we mean by "vacuum" as a class of physical conditions.

Unlike our everyday definition of a vacuum or void, to physicists a vacuum is a very specific condition that can be mathematically specified: It is the state of a physical system in which the system has the lowest possible energy: "You can't have less energy than nothing at all." If you take away the energy in free particles and fields that zip through space, you end up with something that we would easily recognize as the familiar "empty" Void. A physicist would add to this definition by saying that a vacuum is also the state containing virtual particles that are hidden under the cloak of quantum uncertainty and that also yield a net energy of zero. But the Higgs field changes this definition in an important way because it is always there, in the Void, interacting with itself even when nothing else is around, not even the virtual particles required by quantum electrodynamics. You can't shut it off. In mathematics, you can subtract the symbols for all the other fields in nature that are under human control, but you are always left with the Higgs field. The way the sourceless Higgs field interacts with itself also leads to more than one lowest-energy vacuum state, and this state changes as the universe becomes warmer or cooler. When the universe is cold or when you merely lob particles at each other through space at low speed, the Higgs field has very high viscosity like ordinary engine oil. At very high temperatures and during high-energy particle collisions near the critical temperature, the Higgs field loses some of its viscosity and particles begin to lose their

FIGURE 4.2   Like a bridge suspended above a valley floor, our Void may be suspended above a deep chasm. When the "cosmos" falls, however, space may erupt in fountains of matter and fields that will ooze out of the empty Void like moisture condensing out of thin air. (Courtesy of David Dahl)

mass. The idea that the Higgs field gives the vacuum a latent energy is, by far, one of the strangest aspects of the theory, especially if you happen to be an astronomer. In many ways, this idea is even more counterintuitive than the way the Higgs field is supposed to account for the origin of mass. It is also one of the most important new ideas to come along in physics since the discovery of virtual particles. It changes everything. Our entire universe might be suspended above an invisible chasm of still-lower vacuum energy like the bridge in Figure 4.2. Could our physical world, someday, go through an unimaginably catastrophic event as the vacuum collapses upon itself and our cosmic "bridge" comes crashing down?

The first thing that an architect studies when designing a new house is the lay of the landscape and the quality of the ground on top of which the building will be constructed. This can change depending on where you are located on Earth. As you walk around on the surface of Earth, your weight will change because of two factors: the gravitational force and the centrifugal force. The surface of Earth has a gravitational field that depends on the distribution of matter below its surface. Because Earth rotates once every twenty-four hours, you are being gently pulled away from the center of Earth by a slight centrifugal force. The combination of these two opposing forces means that there will be places on Earth where your bathroom scale will register your weight a half-pound lighter or heavier than it will elsewhere. A 200-ton building erected near the equator will weigh a whopping 1,300 pounds less than a similar building built at the poles. It will be slightly taller because the materials are under less compression as the centrifugal force at the equator tries to sling the top of the building into space. So, as an architect, the way you build buildings is affected by invisible forces "in space." After you had built enough of these buildings with the same features, you could actually detect the changing properties of space by measuring, say, the final height or weight of a building.

The way that the Higgs field works is similar to this analogy, only it is the mass (not the weight) of the electron, the quark, and the W and Z particles that serves as the gauge. Now, whenever physicists try to create a mathematical model of how they hope particles and fields are supposed to act, they have to first understand the exact properties of the vacuum state upon which the physical world is built. Still, the properties of the Higgs field are not written in stone. They are intentionally defined in a specific way by the architects of electroweak theory. The Higgs field can be mathematically tailored to produce a unique set of hills and valleys of latent energy as it interacts with itself. This would be like an architect being able to modify the gravity and rotation speed of Earth to get just the right height or weight of a building located anywhere on Earth. This latent energy, thanks to Einstein's theory of relativity and $E = mc^2$, appears as a slight change of mass of the particle with which the Higgs field is interacting. This tailoring process will determine the masses of the quarks and electrons and the particles that transmit the strong, weak, and electromagnetic forces. All of these properties have to be tuned in just the right way to "create" our particular universe of particles and forces, otherwise the whole theoretical building will eventually collapse on its foundations. This is where physicists have to open a dialogue with astronomers. The type of vacuum the physicist starts from not only has to give back the desired physics, it also has to give the same kind of Void that astronomers have been studying for centuries. After all, atomic space seems to be smoothly attached to the space in our labs, and *that* space connects with the space in our solar system, galaxy, and cosmos. It sounds like an altogether confusing process, but fortunately, it is highly interconnected with distinct patterns. Luckily, Nature provided us with a system that has some of the same features as the Higgs field.

Most of us think of magnets as being fixed and resistant to change. Kitchen and toy magnets work in the same way day after day. In fact, they are a lot more malleable than common experience suggests. You can't change a toy magnet's magnetism by putting it in a standard oven on the "bake" setting, but with a blowtorch you might be in for a surprise. If an ordinary bar magnet is heated to temperatures of more than 570° Celsius, it will lose its North and South polarity. This happens because the individual atoms that carry their own magnetic polarities begin to shift slightly within the solid crystalline structure. Their magnetic poles become randomly oriented. If we cooled the magnet below this temperature, the polarity would recover because the atoms would become less mobile and would rearrange themselves in a way that minimizes the local magnetic stresses. In the end, polarity divides the space inside the magnet into two distinct universes. If you didn't have a compass or know about polarity, you would never suspect that as a microscopic denizen, your bar-magnet universe had this kind of structure to it. You would never suspect that there were actually two distinctly different Voids out there with their own polarities, because the (magnetic) field responsible for this separation is hidden in the Void itself. All you would need is the right kind of instrument (a compass) to figure out whether you were living in a North-type or South-type Void. Either way, your universe would look the same in outward appearance. The architect who builds the same building in different parts of the world would hardly notice the fraction-of-an-inch change in the building height caused by the centrifugal flexure due to the rotating Earth, but a physicist might. Luckily, our world isn't nearly as pathological as what you might expect to find inside a magnet.

It is an undisputed fact that the Higgs field must operate in the same way everywhere in our corner of the universe. Electrons passing through the solar system as cosmic rays have the same masses as the ones in our laboratories. The atomic spectra from distant galaxies look like the spectra we see from laboratory gases. Electroweak theory says that this is not surprising. The way the universe looks depends only on where the thermostat for the universe is set with respect to a temperature that is near 1,000 trillion degrees Fahrenheit. For us, the universe is nearly 1,000 trillion times colder than that. Astronomers have measured the temperature of the universe and consistently arrive at a chilling number that is 2.7° Celsius above Absolute Zero. If it had been closer to 1,000 trillion degrees Celsius, we would have had cause to worry. But in different parts of the universe, past the farthest galaxies we can detect, who knows? Space might actually be a patchwork of different voids, each built on a different state of the Higgs field like the pattern of ice patches in a frozen pond. Traveling from one of these patches to another might lead to spectacular changes in the masses of the particles that serve as our compasses. There also might be changes in the number of forces that knit the world together. There may be a void out there somewhere where an electron and its heavier cousin, the muon, have swapped masses. Our proton might be someone else's Neutral Xi particle.

These are all very exciting speculations, but we need to realize that unlike QED, which has been tested extensively, electroweak theory still has a few more hurdles to conquer before we can fully accept the need for something as provocative as a Higgs field. This means that although we should accept what QED has to say about virtual particles in the Void, the bizarre landscape painted by electroweak theory is still based on a theory not fully tested. There is a logical escape clause we can invoke to maintain a skeptical viewpoint and philosophy. After all, the way that the Higgs field operates as an invisible essence lurking in the Void has an uncanny parallel to the older Ether idea. Its tendency for not appearing directly under laboratory conditions only increases the suspicion that we may have gone astray in using one pattern to explain another. The irony is that the 1979 Nobel Prize in physics had already been awarded to Steven Weinberg at the University of Texas, Abdus Salam at the University of Trieste, and Sheldon Glashow at Boston University for their work on electroweak theory. There are no historical precedents for revoking the prize even when the awarded idea falls flat.

If you were to take a survey of how physicists actually felt about the Higgs field, you would not see candid opinions flocking to support this idea like birds following some mindless urge to fly to Capistrano. In some corners of the physics community, if you mention the Higgs field, you will be treated to icy stares, exasperated denials, and even simmering tempers. Steven Weinberg was terribly surprised that his "toy model" for unification actually worked as well as it did. Peter Higgs, a physicist at the University of Edinburg in Scotland and the discoverer of the Higgs mechanism, was disturbed that this field had been named for him. Leon Lederman, the director of Fermilab, wrote in his 1994 book *The God Particle* that this wraith-like field should probably have been called the "goddamn particle" because it would completely stifle any further progress in understanding the true nature of mass, when it is found. Finally, and in a more colorful turn of phrase, Sheldon Glashow often refers to the Higgs field as "a toilet down which we flush away the inconsistencies of our present theories."

As the twenty-first century begins, we find ourselves at a historic moment. A theory has passed all of its tests except for one, and that one is the most critical of all. While experimenters search for the Higgs particle, other physicists have not waited for the final experimental validation of electroweak theory. They have stormed ahead to use it as a pattern to investigate the strong force, and even gravity itself. Over thirty years of theoretical research into other areas of unifying the particles and fields now stands or falls on whether or not the Higgs mechanism is correct. The success or failure of a huge amount of research by thousands of physicists all comes down to one or two crucial questions: Does the Higgs field really exist? Can we pump enough energy into the Void that a virtual Higgs particle will materialize in our detectors as a real, physical particle? If this book had been written in 1998, the rest of this story would have been a chronicle of negative results. But, as the first year of the new millennium drew to a close, there were exciting hints that the answer to even these questions might well be yes. These hints were dashed six months later by a new round of detailed investigation.

After more than a decade of operation, the Large Electron-Positron Collider (LEP I), built by the European nuclear research consortium CERN in Geneva, was going to be shut off in September 2000 to make way for its upgrade to the more powerful LEP II machine by 2005. All of a sudden, scientists began to report something odd going on. With the ALEPH and DELPHI detectors attached to LEP, each as big as a three-story building, scientists had spotted several subatomic collision events (out of millions) that might have been the traces of Higgs particles being created and destroyed. Why hadn't they seen these tantalizing traces during the past decade of operation? It was rather simple. They did exactly what any of us would have done if we knew that a machine we were using was going to be relegated to the junk pile in a few months. In a damn-the-torpedoes move, the LEP scientists had run up the energies of each of the two colliding beams to energies of 103 billion volts, an amount of energy per particle that is equal to converting nearly 103 protons into pure energy. This change pushed the six-mile-long machine above its operating envelope, which was normally set at about 100 billion volts. It was an eleventh-hour move of desperation because in a month, the entire machine would be shut down. If it broke then, it didn't matter. This slight increase by only 3 billion volts may have been just enough to turn disappointment into success.

The two events teased out of the flurry of data in August and September were joined by a new event on October 16, 1999, seen with a different instrument called the L3 detector, and then yet another event in November. The hints of a new particle in the making were captured against the fireball of the electron-positron annihilation. Whatever this mysterious new particle might have been, it promptly decayed into other much less interesting particles, but it left behind a fleeting fingerprint of missing energy that could have been from one of the carriers of the weak force: the Z^0. Because the electroweak theory says that Z^0s are always supposed to tag along with the Higgs particles, the original heavy particle appeared to be the Higgs particle. Based on the information culled from the handful of curious events seen at L3, ALEPH, and DELPH I, the mass of this mysterious particle would have been a staggering 114 billion volts, which is more than the mass of an entire atom of gold. The accelerator had pumped nearly the right amount of energy into the Void so that one of its ghostly Higgs particles became a real particle for just an instant.

Because the conversion of the LEP I to the LEP II machine requires shutting off the LEP I until 2004, a brief battle erupted between the CERN scientists and the much beleaguered director-general, Luciano Maiani, with the scientists begging to delay work on the LEP II for about a year, so that more data could be gathered. The cost would have been several million Swiss francs, a tremendous waste of money if further data did not support the Higgs detection. The plug was finally pulled at 8:00 A.M. on November 2, 2000. Everyone realized the irony of shutting off a machine just as it may have been about to make the biggest discovery in physics in the last twenty years. A firm detection would literally have justified the billion-dollar machine's existence. In some ways, it was like decommissioning a new bridge after the first handful of cars had successfully driven across it. The objective of the bridge was merely to prove that a car can travel on a suspended roadway to the other side, not to supply a permanent means of doing so. Engineers now had to close the bridge and build a new one to see if even heavier eighteen-wheelers could also make the journey. Across the Atlantic from CERN, beginning in March 2001, the upgraded Tevatron accelerator at Fermilab in Batavia, Illinois, became capable of creating 100 times more of these mysterious events than the LEP I had in its final months. Many physicists are convinced that any scientific discovery and accompanying paper for the Higgs particle will probably come from Fermilab. As it turned out, all this excitement over a handful of promising data spikes would not pan out to herald a new particle. By June 2001, the same data had been analyzed using more rigorous statistical tests and simulations, only to show that the events were merely statistical noise. The best that could be said of the Higgs particles was that they were heavier than a gold atom, and we had still not seen the slightest trace of them by 2002.

As the fate of the electroweak theory hangs in the balance, continued theoretical research spanning over three decades has pushed the spirit of unification into other domains as the program started by Maxwell reined in an ever-larger net of possibilities. The strong nuclear force transmitted by gluons was eventually shoehorned into a mathematically precise theory called quantum chromodynamics (QCD) that led to spectacular achievements such as the first calculation of the mass of a proton. It took a specially designed supercomputer a full year to do the calculation and cough up an answer that was accurate to about 10 percent. It was nothing like the ten-decimal-place accuracy of QED, but it was good enough, anyway. Electroweak theory and QCD were bound together into a cookbook called the "standard model," and it seemed to work in accounting for nearly everything physicists had seen in their laboratories prior to 2000. The drawback was that it wasn't really a "unified theory" of the electromagnetic, strong, and weak forces. It wasn't a single theory that could be mathematically sliced to give back each of the three forces separately. There were two dozen adjustable factors that had to be entered by hand in order to reproduce the world seen by physicists in their laboratories. Among all of these fine-tunings, the Higgs field was still something of a huge embarrassment. You could neither predict the mass of the particle that caused it nor account for the twenty-four numbers that specified the characteristics of our physical world. Nearly every physicist agreed that it was a very ugly and unappetizing stew of fundamental constants, not at all what you might expect from a streamlined and unified theory. The adoption of the standard model also institutionalized a very unpleasant trend.

The Void had become something of a garbage dump of hidden fields. So long as these fields stayed hidden, it was okay to add more of them to the Void. Within the Void, particles could travel forward and backward in time, violate the conservation of energy, modify masses, and perhaps even travel faster than light. They could do this as long as none of this capricious behavior ever showed up as a property of any real particle or process you could measure in the lab. Each time physicists found a simpler way to describe how Nature was put together, the Void itself had to change dramatically to make the explanations work. QED gave us virtual particles that filled the Void with their hidden, but important, activity. With the right amount of energy, we can bring these hidden particles into the real world to find that they are identical to the particles that we already know about. With electroweak theory, the Void gained a new ingredient, the Higgs field, which painted the vacuum with a patina of vacuum energy states, each affecting the masses of particles in specific ways. The pattern of tampering with the Void took another turn toward the bizarre with the completion of QCD as a workable theory for the strong force. We now get to think of the Void as some kind of quantum molasses.

Quarks and gluons are confined inside nuclei and particles in a way that implies they are *actively* excluded from moving into the surrounding space out beyond the hazy perimeter of nuclear matter. Space doesn't allow these particles to roam freely throughout the universe. The Void, so far as quarks and gluons are concerned, is like a thick, engulfing morass. With enough energy, you could actually melt this vacuum and watch it form again in a fleeting trillionth of a second. For a brief moment, the molasses effect would slacken and the quarks would respond to a weakening of the strong force as they were pulled apart. Experiments now in progress at the Brookhaven Relativistic Heavy Ion Collider in Long Island, New York, will be able to discover whether the vacuum can be altered in the exact way that QCD theory says it must be. If it can't, then there is some critical flaw with the quark-gluon theory that has not been uncovered yet. If the Void can be melted in the way that QCD suggests, however, we must confront an even newer idea about the Void, with a whole new level of hidden patterns and control over matter from deep within its invisible fabric.

The amazing thing about all these descriptions of the Void provided by QED, QCD, and electroweak theory is that they refer to the *same* Void. The descriptions not only overlap, they interpenetrate. Virtual particles (electrons, positrons, and photons among them) in the QED vacuum occupy the same physical space as the Higgs field and the QCD molasses. The Higgs field affects both the QED and the QCD vacuum because it alters the masses of electrons and quarks as well as photons, the W and Z particles, and gluons. The QED vacuum can become the QCD vacuum if the virtual photons have enough energy to produce gluon-antigluon pairs or quark-antiquark pairs, which can then react to the effects of the QCD molasses.

In the quest to find ever simpler ways of explaining the nongravitational forces, physicists were constantly looking for some way to combine all three descriptions into a single mathematical framework—a single set of symbols that could be manipulated to describe each force separately. The electromagnetic and weak forces had already been unified, or so it was hoped. It seemed that all the physicists needed to do was to find an even more powerful set of symbols and mathematical operations to combine the strong force with electroweak theory. There may actually be only two forces in nature: gravity and something we must now call the electronuclear force, combining electromagnetic, weak, and strong forces together into an unbroken and seamless unity. Taking this next, and very obvious, step in unification has turned out to be a tremendously more difficult task that ultimately led to a bona fide dead end in physics.

There is no simple way to describe what went on in the 1970s as the details of unifying the strong and electroweak forces into a single theory, called "grand unification theory" (GUT), were hammered out. Many trails were followed into the dense thickets of mathematical abstraction. Some led to interesting but physically irrelevant technical discoveries. Other trails seemed to go on and on into the wilderness before finally coming mercifully to a dead end. In all the many ideas that followed, a handful of basic issues seemed to dog physicists relentlessly. None of them had a greater significance and persistence than the multiplicity of domains spawned by even newer families of Higgs fields. If you were going to paint a picture of the real world with these brushes, you had to allow for the texture in the canvas formed by the Higgs fields even before you started to add your first brush stroke. But the new families of Higgs particles revealed their existence at energies beyond anything that electroweak theory had ever anticipated, and far beyond anything physicists had ever imagined.

In a direct parallel with what had been built into the electroweak theory, the new Higgs particles were needed so that the strong force would eventually become distinct from the unified electroweak force as the temperature was lowered. This temperature was trillions of times higher than the threshold needed for bringing on electroweak unification. There was also a new issue raised by the new Higgs particles that became increasingly vexing. The properties of these new Higgs particles were in many ways ad hoc. They were not constrained by direct observation. They seemed to have the character of an arbitrary ingredient added to the Void, and this was a bad sign that many physicists regarded with foreboding. To use a colorful description coined by Glashow, physicists had created a mansion that was now equipped with two toilets. There was also another, more serious, dimension to the "Higgs problem" that began to emerge from the theoretical models for grand unification by the late 1970s.

Physicists discovered that every plausible version of grand unification theory added its own new ingredients to the Void, but these ingredients were now enormous in their scope. When physicists calculated how they would work across the vast scale of the universe, they got a nasty surprise. Astronomers said the cosmic Void had nearly no energy to it at all. However, the calculations by physicists seemed to show that the Void had enough energy to swallow the entire universe into a black hole within a few seconds. This could only mean that grand unification theory was wrong in some serious way. Some physicists tried to create newer versions of GUTs where the vacuum energy could be fine-tuned to nearly zero, but the way to do this seemed about as plausible as finding a pencil balanced on its tip in a hurricane. The incremental approach to studying small pieces of the puzzle of particles and fields had at last fallen apart, although no one realized it at the time. It would take another decade before the flaws could be found and addressed. The solution required nothing less than to create a theory that included gravity right from the start, a "theory of everything." It seemed that Nature would permit no shortcuts to be taken to achieve this goal.

5

GRAVITY'S WEB
Space, Time, and Gravity

Past the heavens, seems so far
Now who will paint the midnight star?
—Enya, "Paint the Sky with Stars"

When I was a student at U.C. Berkeley, I used to spend two weeks out of every summer camping in the High Sierras of Yosemite National Park. Most trips were the usual fare—lovely sixty-mile hikes through granite and pine wildernesses along deserted and dusty alpine trails. My recollections of these trips now run together in a long seamless movie of gentle breezes, humming insects, and shady stands of trees that reached for the stars. But of all my memories of this idyllic time, one stands out with special clarity: June 14, 1974. I had been camping for eight days, hiking the many beautiful trails of Tuolumne Meadows and Vogelsang. I regretted that my thrilling and sublime journey was about to come to its inevitable end in a day or so. The weather was a pleasant 75–80° Fahrenheit during the daytime, just warm enough to dry out the trails from the morning's dew and turn them into dusty footpaths by noontime. On that particular day, I was to hike from Vogelsang to my planned bivouac on the far end of Merced Lake, where the trail would lead me into Little Yosemite Valley the next day. The descent to Merced Lake along carefully designed switchback paths was mostly wooded, but occasional vistas along the

83

way revealed the valley and the lake 1,000 feet below. When I had finally descended to the lake and had begun my hike around to the camping area, I looked back along the trail to the far end of the lake, where the surrounding mountains of exposed granite were spectacularly reflected in the lake waters. The skies in that direction were cloudy, at times menacing, as the warmed, moisture-laden air was forced by the winds to rise from the valley floor and then condense into dense masses of clouds.

As I pitched my tent lakeside at 5:00 P.M., I became aware of the wind rustling in the trees. With the prospect that there could be rain headed my way in the next few hours, I put a plastic tarp over the tent. After the long day on the trail, I set about preparing my spartan dinner with considerable enthusiasm. I was enjoying my dinner and the noisy commotion of the local blue jays when suddenly I heard a most peculiar sound. From the far end of the valley, two deep tones sounded in a chain of alternating notes. It was a sound not unlike the foghorns I had often heard on the Golden Gate Bridge in San Francisco. The tones came about one second apart in a slow but powerful beat, and this alternation continued about fifteen times before finally fading away entirely, never to return. I sat there by the campfire totally speechless. I had done a fair amount of camping in these areas over the years under a variety of conditions, and the High Sierra environment had always been peaceful, undisturbed by noises of anything more complex than mosquitoes, birds, the wind in the trees, or babbling brooks. You get used to hearing natural sounds from identifiable sources, and soon they fade into a background that you often cease to think about as you hike. I had never imagined that under the right conditions, the mountains themselves could sound out with their own voices. The experience was also curiously unsettling. I had absolutely no idea how such pure tones could have been caused by any combination of the natural components of this environment. There were probably strong winds up-canyon where thunderstorms were in progress, but how in the world do you get periodic notes from irregular rocks, trees, and the rushing air?

In ages past, sounds like this would have been thought of as a wind created by a spirit of Nature, perhaps the trumpeting of some mountain nymph or deity. Instead, I found myself dissecting the experience to glean from it some germ of understanding and comprehension in terms of modern scientific sensibilities. I was left with a sound produced by a combination of changes in air pressure, gas dynamics, and venturi flow in some distant canyon. While having dinner, I more accurately studied my recollection of the event, but it telescoped away in another direction and I lost touch with its technical significance. The experience was far more than the sum of its scientific parts. In the years to come, I would finally understand how my brain had handled the experience. It became obvious why it had left its mark so vividly in my memories, in such a conflicted, almost spiritual state.

For every experience we are treated to in life, there are actually two separate minds surveying the scene, extracting meaning from it, and passing judgment. The right hemisphere of my brain had taken in the whole picture and instantly stamped it with an emotional hue. It had been alerted to the incongruity of hearing a coherent sound among the incoherent trees and rocks of a wilderness setting. It flavored the event with anxiety because evolution had trained it to instantly frame every experience as potentially hostile or friendly. While this was all going on, the more logical and analytical left hemisphere had been trying to sort out and explain the experience by applying rules and laws, a task that takes longer than the instant appraisal of the experience my emotions made and are still making even today, some twenty-eight years later. Because no obvious rules or prior experiences applied, I became transfixed by indecision as I sat by the campfire, but I eventually decided there were insufficient data to change my dinner routine and move to a safer location. Logic, however, never got the final say in deciding my next responses. I still recall the sense of foreboding and anxiety that washed through me, which had not quite been dissipated by logical reasoning. As nightfall darkened the scene, I huddled nervously in my sleeping bag. For what seemed an eternity, I dozed and woke up, warily scanning the blackened camping area, searching for some threatening specter that never came, whose trumpet call had sounded in the distant mountains.

The human brain knows nothing of abstract ideas when it tries to weigh experiences, categorizing them as good or bad for our survival.

As students, we sat and fretted over math problems for midterm exams as though the outcome would be a life-or-death result. In the thick soup of ideas and sensory experiences, our right hemisphere treats all stimuli in the only way it knows how. It marks incorporeal ideas with the same emotional vigor that it does direct sensory experiences. A gentle touch can summon indifference, cool calculation, or extreme passion. Even a seemingly sterile equation can elicit a near-sexual response or an experience bordering on the religious. When ideas and states such as "darkness" or "space" find their way into this thought work, they stealthily cause us to react with a full orchestra of responses. For instance, let's have a look at the response evoked when you think about the term "dimension." It is one of those obvious, but utterly mysterious, features of space that is both terribly profound and sublimely trivial.

Just as "wet" is a quality of water, dimension is one of the most elementary qualities of space. The one-dimensional universe of a string is rather boring, perhaps of interest only to an ant or a microbe. The space occupied by a piece of paper is two-dimensional, and you can use this to encode a love sonnet, a painting, or a delicate spider's web. Your home is a set of two-dimensional walls and floors that circumscribe a three-dimensional volume, and this circumscription is central to the freedom that lets you admire a waterfall, build a mighty pyramid to house your mortal remains, or launch a spacecraft to Mars. Dimension is a curious property of something that is otherwise invisible—space. Unlike color or sound, dimension is a quality that actually forces you to move about and experience the physical world in only one possible way. It is this invisible governance of what you are and how your world looks that is awe-inspiring and indescribably mysterious.

Since the 1930s, science fiction stories have speculated about traveling to other dimensions or about the existence of universes coexisting with our own but occupying parallel dimensions. All of these ideas have thrilled me since childhood, filling me with an unmistakable feeling of creepiness. As a trained astronomer, I find even more reasons to react to this idea from both an emotional and an intellectual perspective. Amazingly, the concept of dimension is as old as the ancient Greeks, but only recently has anyone explored just how rich the concept is. Little wonder. The world around you, and your way of describing it, seems to require only three dimensions—hardly enough to be intellectually exciting. Up until the nineteenth century, no one could see any need for adding more dimensions to space when no physical need or data seemed to hint that any others existed. You could walk forward, turn left, and jump up. That was it. The way that our understanding of space grew beyond its more obvious three dimensions into seemingly sterile nothingness didn't happen because we confronted a new experience of Nature. It happened through a series of gradual changes in the way mathematicians viewed the entire field of geometry. No one ever woke up and thought, "Today I am going to study the geometry of empty space and twenty-six dimensions." However, through a series of steps, mathematicians crept along the logical ladder of "ifs and thens" to find themselves in a much bigger universe than they had ever imagined existed.

The watershed moment came in 1854 when German mathematician George Riemann (1826–1866) gave his doctor's candidacy lecture to a mixed group of faculty members and other interested parties. Riemann's big idea, which won him his doctorate, was that it was almost as easy to think about the geometry of space for any number of dimensions as it was to think of Euclid's simple two-dimensional geometry scratched on beach sand or paper. Squares and cubes are only the first two rungs in a ladder of similar geometric forms that stretch one after the other into the abstract space of infinite dimensions. In this scheme, lowly three-dimensional space was only one possibility. Riemann's geometries were the first new way of thinking about space in 2,000 years. One of his work's most profound implications was that you didn't need to draw anything at all in order to explore the properties of triangles and intersecting lines. You only needed to use a little algebra. The elegance of studying space by using mathematics rather than scratching figures in sand à la Euclid was a powerful, liberating approach to geometry that would eventually become the key to exploring physical space itself. It only takes one more algebraic symbol to add a fourth dimension to space, and to create the formula for a line in such a space. Try as you might, you can never construct such a fourth way by drawing lines on paper. Also, there was another implication of this new way of studying space: It didn't matter that the physical act of drawing in the sand led to one set of ideas about geometry and working with abstract mathematical symbols led to another. For brains designed to blend experiences and ideas together, mathematical experiences and patterns can be just as compelling as real physical experiences presented to the mind by the senses. Suddenly, the beauty that geometers see with their eyes as they draw lines on paper becomes the same beauty that mathematicians "see" in a set of equations describing the same geometric relationships. Then came the inevitable mental reach from the world of abstract symbols to concrete reality. Physical space might be more complicated than anyone had ever imagined.

Although Riemann originally dealt with abstract space, his mathematical discoveries made it almost irresistible to think that our physical space might be just as complicated. Newton's physics, with its refinement in the eighteenth century, gave physicists the tools they needed to account for the movement of matter in great detail. With Riemann, a new idea blossomed that led to an extraordinary expectation: If Euclid's flat, three-dimensional space worked so well to define the underpinnings of such far-flung ideas as Newton's physics and land surveying, why shouldn't there be a role in the physical world for Riemann's four-, five-, and even infinite-dimensional geometry? Armed with the powerful microscope of his mathematics, Riemann actually explored the properties of physical space itself.

Physical space, to the extent that we can think of it at all, is only one specific type of space that mathematicians work with, though it isn't even particularly interesting to many of them. They really don't know what their mathematics is supposed to represent in the physical world, for the simple reason that geometric ideas don't have tags on them to identify which worldly things they are supposed to represent. When mathematicians speak of distance, it is a pure number that could just as easily mean the distance between two points in miles or the distance between two economic states in U.S. dollars. Economists define the economic state of the world in terms of economic dimensions (consumer price index, unemployment, inflation, and so on) that define a space every bit as real as physical space is to physicists.

Entire national economies move about in this space in accordance with a set of specific economic rules. Space, in whatever guise it takes, isn't a "thing" so much as a pure intellectual construct we use for various kinds of bookkeeping. Its qualities depend only on what it is we are trying to describe at a particular moment. For a physicist, even the physical space you navigate from instant to instant isn't as well-defined as you might imagine.

Despite its familiarity, pragmatic physicists don't subscribe to the kind of space that our common intuition demands. Albert Einstein didn't think about space as something with a natural set of properties like "water is wet" or "ice is cold." In whatever way we define space, our definition always depends upon a reference to material things such as rulers or matter. Einstein's quote in *Relativity: The Special and General Theory* is rather specific about this point: "It is not clear what is to be understood here by 'position' and 'space' . . . In the first place, we entirely shun the vague word 'space,' of which, we must honestly acknowledge, we cannot form the slightest conception, and we replace it by 'motion relative to a [system of coordinates].'"

Look around you a moment. You will see just how convincing the existence of space is. If space didn't exist, everything would be slammed up against your face. Instead, it is obvious to our brains and the way they dissect and reconstruct the world, that things are definitely "out there." Einstein wasn't denying any of this; all he was saying was that there is no way to describe space by itself, except in relation to some physical event or meter stick. Space has no natural properties as Newton required, nor does it affect matter in any way as a medium.

The second thing that is important to consider is that when we talk about the geometry of space, we mean the patterns formed by the things that are embedded in this space like light rays, string, or meter sticks and what their shapes tell us. We are not directly referring to the properties of space itself. Matter provides the anchor points between which lines are drawn or light rays pass. The shape of these light rays or the motion of the clumps of matter then tells you something about the space in which they are embedded. You never see this space directly. The geometry of space just refers to the pattern created by the matter and light rays, not about some road-way on which they are moving. The next time you make spaghetti, look at the way it tangles in the colander. In physics, light rays and matter are like the pieces of spaghetti in an invisible colander. From the way the spaghetti is shaped, we can learn something about the geometry of the invisible colander, just as exploring the paths of light rays can tell us something about the geometry of the space they move through. But it isn't the background space of the colander that actually constrains real particles. It is the spaghetti itself that defines its own space.

Riemann never really said much about the shape of space, just as Newton before him didn't want to speculate about gravity. Was Riemann's mysterious space supposed to represent the Ether, which was still in vogue among physicists at the time, or was it some new underlying, invisible bedrock in which the other things were embedded? On February 21, 1870, before the Cambridge Philosophical Society, the twenty-five-year-old mathematician William Clifford (1845–1879) presented a paper entitled "On the Space Theory of Matter," which took Riemann's ideas about the shape of space on an intellectual joyride. Riemann had been dead for four years, but even he would have been impressed with how far the young Clifford had taken his ideas. Clifford wondered whether it was possible that microscopic pieces of space could be very lumpy and only appear smooth in an average sense. The physical world might then be nothing more than these distortions in space piled one atop the other. Newton's speculations that matter may simply be a condensation of the essence that causes gravitation had now been independently discovered by Clifford. But exactly what was being deformed? Clifford and Riemann thought geometry itself was in many ways a physical thing like air or water. Since neither of them were physicists, and worse yet, they were trapped in the wrong century, they could not make the necessary connections between abstract space geometry and the physical world that would have been compelling to physicists at that time. But even if they had succeeded, no one would have taken their speculations very seriously. They were, after all, not physicists. Another reason speculations of this kind did not fare very well is that they had the misfortune of feeding into a much larger and growing public debate about matters far beyond the confines of pure mathematical research. Surprisingly enough, it wasn't just mathematicians who spent long hours exploring other dimensions to space. Somehow, a much wider audience had engaged these ideas by the early 1870s.

While mathematicians were trying to understand hyper-dimensional geometry, other people impatiently took the exploration of this new landscape into their own hands. The May 1, 1873, issue of the journal *Nature* is interesting, not only for Clifford's English translation of Riemann's groundbreaking 1854 lecture but because it contains a curious article, "On Space of Four Dimensions," by the English philosopher G. F. Rodwell. He mentions that among others, Clifford "had indicated in some of [his] profoundest mathematical demonstrations that [he] possessed, 'an inner assurance of the reality of [four-dimensional] space.'" Could the way we experience space have something to do with our state of motion? Rodwell was convinced that as we eventually learned more about the distant universe, this arena would "lead us to the very threshold of transcendental space, and once on the threshold, we may look wonderingly beyond." Other articles in this issue of *Nature* such as "On Venomous Caterpillars" resulted in several letters to the editor in later issues; surprisingly, Rodwell's and Clifford's articles didn't make a single ripple in the letters from the readers of *Nature*. But then the unexpected happened.

Astronomer Johan Zollner from the University of Leipzig wrote a book in 1878 called *Transcendental Physics*, claiming to have concrete proof that a fourth dimension not only existed but was inhabited by spirits. This led to a curious relationship with the American medium Henry Slade, who was busy giving séances and raking in large sums of money based on public statements along these same lines. These claims got him into serious legal trouble in 1878, when he was hauled before the courts on charges of operating fraudulent séances (is there any other kind?). This opened up the entire subject of the fourth dimension to public inspection, but sadly, it also welded the otherwise abstract mathematical concept to the most radical ideas about the physical world. In an impressive array, physicists, including Zollner himself, were paraded before the courts and the general public to decide whether Slade was deceiving people about actually contacting their departed relatives in the fourth dimension. It is hard for us to imagine today how gripping this public debate might have been at the time. When they weren't busy having legendary mummy-unwrapping parties, Victorian high society apparently followed this bizarre story for months on end as a favorite parlor-room debate. This, however, was not the end of the road for arcane speculation about the dimensions to space.

As the nineteenth century drew to a close, even more incredible theories about the fourth dimension came out of the woodwork, including the proposal that the Ether is a four-dimensional fluid that sweeps through our three-dimensional space. Atoms were the entryways into our universe of the Ether. Philosopher Charles Hinton also wrote a short article, "What Is the Fourth Dimension?" in 1887, a delightful description in intuitively appealing prose of how four-dimensional objects would appear as solid bodies in our three-dimensional world. Toward the end, he offered a rather remarkable proposition: Where else would nature choose to manifest its four-dimensional character but at the atomic scale? Atoms might actually be thin, four-dimensional threads whose cross-sections are seen as microscopic, three-dimensional bodies. You can imagine that amid all the popular speculation about the fourth dimension, physicists may have been a little reluctant to give the subject much serious thought. In fact, the level of disdain for this subject was so acute that the major scientific journals of the time seemed to avoid any mention of the fourth dimension at all. Amid all the careful mathematical advances rapidly being made in studying "N-dimensional Riemannian geometry" and the upsurge in interest in ghosts and spiritualism, the German mathematician Hermann Minkowski (1864–1909) emerged. He was soon to play a pivotal role in creating an entirely new language for describing physical space. We don't really know whether Minkowski's readings as a young man included the popular literature of this period, when spirits and the Ether hailed from the fourth dimension. But we do know that toward the end of his life, he revolutionized an obscure corner of theoretical physics called special relativity, transforming abstract discussions about the fourth dimension into the basic currency of twentieth-century physics.

At the age of thirty-one, Minkowski had the rather minor notoriety of being Albert Einstein's math teacher at the Zurich Polytechnic in 1895. This, by the way, was the same year that H. G. Wells (1866–1946) published *The Time Machine*, which dramatically proposed that the fourth dimension was not a spacelike thing but just another name for time itself. We don't know whether Minkowski ever read H. G. Wells, or exactly where Wells got his idea, for that matter, but Minkowski seemed to have become acutely aware of the idea at some early point in his life. Many modern authors often imply that Einstein somehow discovered the fourth dimension. In fact, it was actually Minkowski who uncovered its most appealing properties. Einstein treated space and time in his mathematics of special relativity as separate features of existence, while at some point in his study of physics, Minkowski saw them as a blended, single object. Minkowski has often been nothing more than a footnote in the development of special relativity, a theoretical comma tacked to the end of Einstein's great work. In many textbooks that discuss the technical aspects of special relativity, Minkowski isn't even mentioned at all. But the very language we now use in physics to describe space and time is derived from Minkowski's geometry, not Einstein's algebra. Long before Einstein learned about geometric forms called tensors and their use in physics, Minkowski had already applied them to electromagnetism, uncovering a beautifully simple way to describe much of the physics known by the turn of the twentieth century and to explain it all in terms of purely geometric ideas. There was, however, a deeper mystery to space that Minkowski was able to lay bare because of his unique experiences with geometry and Riemann's way of thinking.

Minkowski realized that the true arena for Einstein's theory of relativity, published in 1905, was not three-dimensional space with a dangling time dimension tacked on as an afterthought. Instead, time and space were equal partners in a four-dimensional landscape that had its own geometric rules, much as a flat sheet of paper must slavishly call forth all of Euclid's geometric postulates. Space and time form a single, continuous mathematical object—a new kind of hybrid space that seamlessly combines the attributes of space and time together. Like a true mathematician freed from the worries of physical intuition, Minkowski saw every event in Nature as a collection of four numbers, or "coordinates." It didn't matter that three of them were space and one of them was time. All that matters is that sets containing four numbers define some kind of abstract space where every point has a unique address and where there is a prescription for calculating the distances between these points based on their addresses. For example, knowing that houses have consecutive numerical addresses, and with two simple rules, you can figure out whether a house is on the left or right side of a street. To get to a different address, you can determine which direction to go and how far you have to travel to get there. All that Minkowski saw was a valid space of four dimensions, and that this space had to follow a specific kind of geometry in order for it to represent the physical world. As Minkowski said:

> The views of space and time which I wish to lay before you have sprung from the soil of experimental physics, and therein lies their strength. They are radical. Henceforth space by itself, and time by itself, are doomed to fade away into mere shadows, and only a kind of union of the two will preserve an independent reality.

You might have thought Einstein would have been excited about Minkowski's breakthrough in reinterpreting special relativity. He wasn't. Einstein wasn't very impressed by Minkowski's geometrical way of describing space-time. In fact, there are some indications that he was actually pretty annoyed by it. He even said that Minkowski's geometry and tensors were nothing more than "superfluous learnedness." Einstein eventually adopted many of Minkowski's methods— nearly a decade later, in 1912—when he began work on ideas bigger than special relativity. Sadly, this acceptance came too late for Minkowski. He never lived to see his geometric approach to relativity become its standard currency. Minkowski died in 1909, well before Einstein published his theory of general relativity, which essentially brought Minkowski's geometric language into the mainstream of physics. It is easy to see why Minkowski's ideas had such a hard time catching on. They presented very difficult and unfamiliar ways of thinking about physical phenomena. There was a whole raft of new terms to master: "space-time," "geodesic," "worldline," "metric." In the end, the reward for all this effort seemed rather modest.

Accepting Minkowski's space-time approach to physics meant following him on a nearly impossible journey. He was asking that we use the logical language that our left hemisphere prefers to use to sort out cause and effect and events in time and apply them to describing the spacelike patterns that our right hemisphere prefers to use. In essence, he wanted us to think with both sides of the brain at the same time. Our brains have been tuned to let the left hemisphere do all the talking and organizing of things in time, while our right hemisphere flavors ideas with emotions and in the context of the big picture. Our right hemisphere is probably more adept at visualizing space-time than our left hemisphere, which constantly tries to deconstruct it into logical, cause-and-effect sequences of events. Our brains haven't prepared us to force the right hemisphere to use the logic of the left. In fact, the right hemisphere is practically mathematically illiterate. It is also inarticulate, so that even when we do manage to succeed at appreciating space-time as a single object, we cannot consciously explain it to ourselves in words.

The problem with our intuitive way of looking at the universe and the events that take place in it is that Nature doesn't really let you separate time and space into artificially distinct compartments the way our conscious, talkative left-hemisphere mind likes to. According to special relativity, statements about time and space experienced by a person moving at high speed become badly mixed together as seen by someone else. The reason Newton's physics doesn't correctly describe high-speed events is that it is based on a misinterpretation of what space and time are. It doesn't account for how maddeningly slippery it is to confine time and space to separate domains of existence. To see how it is sometimes hard to go from perceiving separate ingredients to experiencing a complete whole, let's take an example of two physical processes we have experienced many times: electricity and magnetism.

Electricity and magnetism are very distinct phenomena. You get electricity from a battery or a wall socket, and you get magnetism from a magnet. There is one similarity between them: They both have polarities. Magnetism's North and South poles are reflected in the positive and negative charges that carry electricity. Maxwell's electromagnetics and Einstein's relativity show us that electric and magnetic fields are not separate things at all. How we experience them depends on our particular frame of reference. A moving electric field looks like a pure electric field to a person moving with the current. It looks like a magnetic field to someone "at rest," not going with the flow. The same underlying phenomenon can look like two very different physical processes merely because of the point of view of the observer. The surprising thing about relativity is that it explains how the experience of time and space can be just as pliable. For things moving slowly, Newton had trained us well, and our left hemispheres are able to disentangle a unique and reproducible time-ordering of events operating in space. But twentieth-century experiments had already begun to explore very extreme environments in Nature. Rough edges were beginning to show up, and only Einstein's relativity with its mixed-up time and space ideas served as a familiar anchor to this topsy-turvy world. To understand it, the only recourse was to look at some experiments from a bigger perspective—the space-time perspective. You have to try to see things from an all-at-once vantage point, not unlike the one that our right hemispheres prefer to use.

Despite its often bizarre history, the four-dimensional thinking you need to use to make sense of space-time is not as odd as it seems. It's just another way of looking at the things in the world that are already familiar. Your history is a series of events in your life that unfold in both space and time (what Minkowski would have called a "worldline"). Beginning when you were born, your history worldline ends with your death. Let's imagine that for every minute of your life up until now, someone had kept detailed notes about your latitude, longitude, and altitude. A computer could take these millions of pieces of data and build a four-dimensional plot of your worldline from birth to the present moment. Each point on the line would represent your position in space and the time you got there. But the line would be incomplete because to be a true worldline, it has to extend all the way to the moment of your death. This information is not available to you right now, but if you happened to be a planet or dust mote, scientists could easily fill in the missing details of your future, using the relentless laws of Nature that controlled you. Now that you have drawn this complete worldline, you can begin to glimpse another mode of thinking. You can see the history as an all-at-once object, not as a succession of distinct "nows." A physicist would no longer describe the forces acting at each moment to explain a specific "now" along the worldline. A physicist would study instead the entire shape of the worldline and perceive the geometry of the force that causes the worldline to have its particular shape, like a trail winding its way up and down a sinuous canyon. Both descriptions are equivalent, but for a particular process, one of these may be easier to use than the other and may lead to deeper insights.

To think in terms of the four-dimensional, space-time frame of reference, you will need to accept that space-time isn't an arena in which things happen. It's an arena where things are present at all instants in their history. In this way of looking at the physical world, your entire worldline *already exists,* even though you have not yet experienced all of it. It is the kind of holistic picture that our right hemispheres are especially well-tuned to digest. If you happened to be a piece of matter, or an entire universe, controlled by the laws of physics and not human volition, we could calculate your future worldline with amazing accuracy, even though these future events are hidden from us today. Furthermore, a worldline can either be drawn on a flat piece of paper or on a curved one. In the latter case, this curved geometry makes a very big difference in how things move.

Driving along one of those straight stretches of highway in Nevada or Kansas, we are barely conscious of our cars moving at sixty-five miles per hour with the distant countryside slowly moving past us. We scarcely have to move our steering wheels to keep going in a straight line. From the air, our course is also ruler-straight. While we are driving down the road parallel to its centerline, the roadway passes up and over a small hill. Our car is still traveling in a perfectly straight path, and the roadway carries our car fifty feet above the previous roadway level. There is suddenly a gentle pressure in our seats as our bodies are accelerated downward as the car climbs to the crest of the hill, and upward as we descend to the bottom of the hill. From the air directly overhead, someone else would swear the car never deviated from a straight-line path. The hillock is invisible from high altitudes. Nevertheless, the changing lay of the landscape, in particular the bending of its flat geometry by the hill, caused us to experience a changing force inside the car. If the windows had been blacked out, we would never have known about the hill, or even our motion at sixty-five miles an hour. Only the coming and going of a physical force would have been experienced.

We all learn about gravity long before our first teacher makes us memorize seemingly random facts about it in grade school. By the time we have taken our first steps, our body and mind have learned to unconsciously adjust to this pesky invisible force that holds us to the ground. We soon learn that we can't jump very high, that stepping through an open window is a really bad idea, and that walking is always a process just a stumble away from major injury. But somewhere along the way we also learn that this invisible force is not an entirely random factor in our lives. It follows basic rules that teach us how to get around in our world without injury. Gravity points the way. It makes baseballs or cannonballs travel along similar paths no matter where we are on the surface of the Earth. In 1919, there were only two "theories of gravity" that physicists knew. Newton's physics, well known to everyone, had ushered in two centuries of dazzling discoveries. Einstein's general relativity, however, was known only to a few physicists. What made Einstein's view so amazing was that he saw gravity and geometry as really the same things, described in different languages. General relativity gave physicists a Rosetta stone to translate between the two ways of looking at gravity, but did it really work?

The difference between Newton and Einstein came down to a single prediction about how strongly a ray of light grazing the sun would be bent. With a little bit of effort, it was possible to calculate how much deflection a ray of light would experience as it passed the limb of the Sun on its way to Earth. But the calculation didn't come naturally. Until Einstein had given Newton a rough shove, no one had ever bothered to do the calculation. Newton's mechanics was well suited to billiard balls or planets; it had nothing to say about the behavior of a will-o'-the-wisp-like light. Meanwhile, with general relativity, Einstein had assembled a new way of looking at gravity and motion that incorporated a geometric view of physics and provided a central role for space-time. Hardly anyone knew about general relativity. It didn't help at all that the theory was couched in a very difficult kind of math called "tensors." It put off thousands of would-be explorers, who found concepts such as "four-dimensional geometry" or "geodesic curvature" incredibly intimidating. The bottom line was that Einstein's theory explained what light would do as it moved through space. The theory could extract from the mathematics a prediction for how gravity would bend light, but it was a prediction exactly twice as large as anything Newton's physics could offer. All eyes were trained on the outcome of a total solar eclipse on May 29, 1919, that would settle the dispute. Would Newton again step forward with the winning answer, or would the victory this time turn toward a new generation of thinkers?

On that particular day, the Moon did much more than darken the Sun. When the measurements of the star positions were finally analyzed, Einstein's new theory emerged victorious, thoroughly eclipsing its 250-year-old competition. Newton's physics could, under pressure, account for the deflection, but it came up short in calculating just how big the deflection would be. The deflection seen by astronomers really did prove that the Sun's gravity acted like a lens to deflect the light rays from the distant stars. But there was a second phenomenon that made its own roughly equal contribution beyond what Newton would have predicted. Not only had the Sun's gravity bent the starlight into speedy cosmic bullets, but the Sun's gravity had literally warped the entire volume of space around the Sun. It would take both Newton's gravitation and Einstein's warped space to confirm the predictions made by the observations. Newspapers around the world went wild with this discovery. Until the *London Times* published the findings of the eclipse expedition on November 7, no one had ever heard of Albert Einstein in the popular press. But a headline that read "Revolution in Science: New Theory of the Universe: Newton's Ideas Overthrown" was enough to turn him into an international celebrity. Predictably, the news reports focused on the bizarre world of warped space and the magic of how nothingness could be deformed.

With the success of the 1919 eclipse observations, the rest of Einstein's theories were eventually accepted, one after another. Physicists began to look seriously at other predictions allowed by this new theory for space, time, and gravity to see whether they could be confirmed in the real world. Its confirmation can be seen in the light from distant galaxies shown in Figure 5.1 Amazingly enough, considering just how important this theory has become, it is a very short list of only six predictions. Their eventual confirmation by physicists during the last half of the twentieth century has given us a glimpse of a larger universe that is every bit as mysterious as anything that appears in a science fiction story. However, this world is quite firmly real, not a product of our wishful thinking or fantasy. Beyond the miracle of bending light, general relativity predicted many other things that gravity could do, which were eventually confirmed by careful study. Gravity has an effect upon clocks and light waves, causing clocks to run slower and light to lose energy as it escapes from a gravity field. When you jiggle a body, it creates waves in the gravitational field that carry off energy, much as jiggling an electron causes electromagnetic waves. But gravity waves are waves of geometric distortion in the very fabric of space itself. Close to a rotating body, there is also another subtle effect that is similar to a warping of space; in this case it can be thought of as a warping of time. No matter what you try to do, the space around such a body prevents anything from orbiting it in a stable path with a fixed orientation in space, like a dancer with a permanent case of vertigo. A gyroscope will slowly shift its axis over time. On the cosmic scale, space loses all of its rigidity and actually stretches like a piece of taffy, dragging galaxies and stars with it. We have the paradox of movement without there being any motion through space.

Since 1919, countless articles and books have been written about general relativity and its magical world, and not one of them fails to mention the mystery of warped space. It always confronts us with ex-

FIGURE 5.1   Gravity can bend space and act as a lens through which light can be distorted as though in a fun-house mirror. A distant cluster of galaxies called Abell 2218 reproduces the images of still more distant galaxies, distorting them into arcs of light. Like a cosmic mirage, these arcs do not exist with the same solidity as a star but change in dramatic ways, depending on the vantage point of the cosmic observer. (Courtesy NASA/STScI)

actly the same conundrum: How can the nothingness of empty space be bent? We have taken one big step in wondering how space can have a property called dimension. Now we have to ponder the equally enigmatic question of how the same empty nothingness can be bent and stretched. As Einstein himself noted in his book *Relativity: The Special and General Theory*, "The non-mathematician is seized by a mysterious shuddering when he hears of 'four-dimensional' things, by a feeling not unlike that awakened by thoughts of the occult." And "seized" we all still remain.

I learned about the fourth dimension not from serious reading in physics books but from an episode of *Twilight Zone* I watched when I was eight years old. In the story, a little girl rolled off her bed and tumbled through an invisible opening into another dimension. Her father frantically searched through this hazy, shifting landscape until he found her. They made it back through into our world just before the opening began to close again. I could not sleep in my own room that night. I slept in my parents' bedroom on a cot rather than risk falling into another dimension and being lost forever. By the time this theme reappeared on the *Outer Limits* TV show three years later, I felt that this whole business of space's three dimensions was altogether too unimaginative. It would be a far more interesting world if there were more than the three obvious ones I kept running into every day. Wouldn't it be neat if space were some kind of fabric that could be ripped or twisted so that you could move from place to place or across hidden dimensions? I could never understand what it was that prevented such options. What was it that would prevent me from turning a fourth corner on the way to the local market and suddenly ending up in a parallel fourth dimension? In time, I graduated from high school and entered college, embarking on a very different journey into the mysterious landscape of space-time. It was a journey more spectacular than any science fiction story.

On March 23, 1973, at 4:30 P.M., I made a fantastic discovery: general relativity. It came fifty-eight years too late for the advancement of science, but I didn't care a whit. I was sitting in my parents' dining room after having done my homework for my junior-year physics classes at U.C. Berkeley, so I turned to my latest passion: tensor analysis. There were no such courses for undergraduates at Berkeley, but during the previous five months I had managed to teach myself the methods of this exotic form of mathematics. For years, I had heard about tensors. I remember being told that they stood in relation to calculus the way calculus stood in relation to arithmetic. In my college courses I had mastered vector differential calculus, and the lure of understanding tensors presented an attractive challenge. Besides, I knew that I would never fully understand cosmology or black holes without tackling this subject. A book on tensors I bought at Cody's Bookstore on Telegraph Avenue was all I needed to start my private studies. For the next five months, I busied myself by systematically solving problem after problem in elementary tensor calculus. I learned about covariant and contravariant forms, geodesic equations, metrics, and curvature. After working with the mathematics for a while, I was stupefied by how simple it all seemed. All the rumors I'd heard about the difficulty of the subject were valid for high-school students, but not for students of advanced math in college. It was one of the most stimulating moments of personal discovery in my entire life, even to the present time.

When I first made contact with the strange universe of Einstein's relativity as a child, I also felt that a glimpse of some mysterious world lay just around the corner, somewhere down the dusty trail of my life. Space-time, worldline, curvature, gravity—these were all terms I'd grown familiar with since I was fourteen years old, the rambunctious outgrowth of a passion for reading science fiction. In junior high school, I taught myself about the weird world of special relativity and the Alice in Wonderland effects of near-light motion. My good friend Renden Holloway and I would calculate length-contraction or time-dilation effects in the back of the classroom while the instructor discussed snail reproduction. We read science fiction all the time and passed notes back and forth in class that sketched mythic space battles. Not a few of them were confiscated by a disapproving science teacher. I especially enjoyed trying to decipher cryptic, spine-tingling phrases like the one in James Blish's 1959 novel, *Galactic Cluster:* "I've heard the commander of the world line cruiser traveling from 8873 to 8704 along the world line of the planet Hathshepa which circles a star on the rim of NGC 4725, calling for help across eleven million light years, but what kind of help he was calling for is beyond my comprehension."

When I finally returned to these ideas as a college student six years later, it made all the difference in the world. I was armed with more mathematical maturity. The impact was nothing less than mind-boggling.

At the same time I was learning how to use the mathematical tools of the tensor calculus, I kept my search physically grounded by reading Einstein's general relativity theory, which was the core of modern cosmology and the then-young subject of black hole research. It didn't take long before I was uncovering the mysterious properties of black holes and big bang cosmology, just by applying the mathematical tools I had mastered. For the first time, and by my own hand, I retraced many of the trails taken by Einstein and other great thinkers as they explored the mysteries of gravity. The mathematical symbols were beautiful curlicues, Greek letters, and brackets that looked lovely when written carefully upon a fresh, clean page. I ground through the tedious calculations that went on for page after page of abstract symbols, ending with three well-formed predictions for how gravity would affect matter and light and explaining why Newton had been left in the dust. I was filled with awe and excitement, more intense than anything I had ever experienced in studying science before. It didn't matter that I wasn't the first one to see these glorious vistas or that my discoveries had happened decades too late to matter to the history of science. For weeks, my waking world seemed to float in some altered state. The emotional impact of these ideas to a receptive young mind was so overpowering that I had to tell everyone I knew. I had to write extensive personal thoughts about it in my diary. For a short while, I imagined what it must have been like for Einstein himself to have wandered through this same territory as a lone pioneer in 1914, scaling these same mountains for the first time in human history. I finally grasped what these terms meant in their full mathematical glory. In the end, I could understand in my own way that everything in the cosmos had something to do with the eternal mystery of gravity. What is gravity? How does it work?

And then came the letdown.

I began to realize that the mathematics I had just mastered held no clues to the answers to these questions. They didn't even help me form a single speculation. Like a latter-day Newton, I could now study black holes, but I was no closer to understanding what gravity *really* was than Newton had been 300 years earlier. Then again, you could read all of Einstein's writings, only to discover that he didn't have any great ideas about gravity, either. He could describe how it worked, not what it was. What is the relationship, after all, between space-time and gravity? Can we, for example, shut off a gravitational field and leave a perfect vacuum behind that is completely empty? The surprising answer is no. As Einstein himself told us: "Space-time does not claim existence on its own but only as a structural quality of the [gravitational] field." This is one of the most far-reaching and mysterious ideas in all of modern physics, and one of its best-kept secrets. It was the centerpiece idea I had been searching for since the time I first tangled with general relativity as a college student. It took twenty years before I stumbled upon this simple quotation, and again I experienced an intense rush of emotions and feeling of awe, this time at the age of forty-one. What was ultimately so humorous about my latter-day revelation was that part of the answer to what gravity was had been staring me in the face all along.

General relativity doesn't hide from us its prescription for thinking about space-time. It is not supposed to be something aloof or abstract with the gravity plopped on top of it like frosting on a cake. Instead, space-time is another name for the gravity itself. The symbols that represent gravity and space-time are not two separate ideas represented by two separate symbols. They are represented by only a single symbol in the equations. This means they are logically identical and indistinguishable. There is nothing else. If gravity could be shut off, you would not end up with a perfect vacuum in space-time; space-time itself would be annihilated. This seems like such an absurd idea that you immediately question it. After all, electromagnetism, like gravity, is also a field, but we can turn it on and off with a light switch without running the risk of destroying the universe. As I read through the early history of general relativity, I discovered that some physicists found this such a bizarre possibility that they immediately began to hunt for another kind of theory to replace Einstein's version. They worked very hard to create competing theories to Einstein's that would break the connection between gravity and space-time. These competing ideas were called "prior-geometry" because they all required that the geometry of space-time had to come first and then the gravitational field would be added on top of it. The problem was that none of them worked.

Prior-geometry theory sees Einstein's space-time as actually a compound object in disguise; one part is the ordinary gravitational field, the other part is the preexisting, immutable space-time. Can such a decomposition really work? The answer is a firm no. The paradox that results from this kind of splitting is that if matter affects the gravitational field and gravity is supported by prior-geometry, then prior-geometry has to be affected by matter even in a roundabout fashion. This violates the entire spirit of prior-geometry in the first place, because now some of its actions are logically indistinguishable from those of gravity. The most important failure of this whole train of thought is that no observation in Einstein's time or since has ever uncovered any physical evidence to support prior-geometry. This means that if we are looking for the source of the mysterious attributes of "geometry" and "dimension" to space-time, we have to look at the gravitational field itself. Imagine how frustrated an intelligent fish would be trying to understand what water is. Ignoring for a moment the occasional fixed reference points of the nearby reef or ocean bottom, there would be no markers to guide the fish to invent a fixed geometry. Its universe would be completely contained within this watery medium. Any equations the fish scientist creates to describe its universe would only tell about the water, not the air above its surface or the rocky basin below. Because there are no solid features to the fish's fluid world, it wouldn't have much use for a concept such as "coordinate" or "point." The fish never needs to ask how it is that water provides from within itself concepts like these. They aren't necessary for its picture of the world. It is a fish, after all. Beyond the meaning of dimension and geometry in the description of space-time and the gravitational field in our world, however, something still more primitive lurks.

If we follow along with most physicists, even the four coordinates of time and space that we use so freely must also in some sense be an intimate part of the gravitational field. There is no background prior-geometry to supply them. This is a subtle but disturbing way of thinking about how we experience space and time. For a moment, the mind draws a complete blank about what to make of this realization. At first, there seems to be nothing about time and space that could be coded in the gravitational field. Time and space themselves act like a kind of scaffolding that all observers use as a marking device wherever they find themselves, just to keep track of other physical parameters such as the strength of a field or the state of matter. In this respect, coordinates like latitude, longitude, altitude, and time have a lot in common with ordinary street addresses, which are physically irrelevant unless there are houses in place first to give them meaning. The city lights of North America seen in Figure 5.2 are an example of

FIGURE 5.2　Many authors describe space-time as though it were a network or fabric of events connected by their histories like a spider's web. Physicists think of space-time as only this network of paths of particles and their points of interaction. Nothing else between these lines and points really exists. City lights in a satellite photo of North America show the same pattern of filled emptiness. (Courtesy DMSP)

special "coordinates" singled out in the darkness, much as we single out specific points in space and time. For a fish, these coordinates are like the knots in a fisherman's net that let the fish know where it is, but there are an infinite number of ways the fisherman could have tossed his net into the sea to surround the fish.

I have tried very hard as an astronomer to figure out what the terms "space-time," "dimension," and "coordinate" mean in a physical way. I almost always end up badly confused, or at least a bit shaken. It isn't because I can't use them properly in the mathematics. The stumbling block that I encounter time and again is that I find it very hard to believe that something as insubstantial as a gravitational field actually causes space and time to exist and make these particular terms *meaningful.* A mountain made of rock allows you only a limited number of ways to hike to its summit. How does the invisible gravitational field force us to behave in only a limited number of ways? We can't turn through a fifth or sixth dimension, no matter how we spin and gyrate.

I am completely incredulous that something as insubstantial as gravity can provide me with the local sense of dimensional space and time. A part of me quietly yearns for the strangely more intuitive idea of prior-geometry, which, like the Ether, promises to set everything upon firmer supports, there behind the draperies of the gravitational field. But I realize that this doesn't make sense, either. I am rebelling against gravity simply because I can't for the moment see how it forces anything to be the way it is. My chauvinism of favoring walls, floors, or clocks to bring solidity to my three-dimensional world seems an unavoidable bias I can't escape. I would have a much easier time of it if I were a jellyfish in water. Is there some way to cleave gravity into its time and space components by analogy with another system? I turn back to Maxwell's electromagnetism after Einstein had worked it over in his special theory of relativity. Armed with these tools, I try to find comfort in their simpler and more intuitive mathematics.

Long ago, in a cramped lecture hall, my physics professor once told his class that you can always describe the electromagnetic field of a particle in terms of its separate electric and magnetic components. The problem is that no two people who are moving with respect to the system that is generating these fields will agree exactly on how much of the field is in its electric or in its magnetic state. But, as with time and space, we all believe we can distinguish between electric and magnetic effects. After all, charged bodies are attracted to each other by electric fields, like a rubbed balloon clinging to a wall. No magnetic attraction there. Magnets, on the other hand, behave very differently. They don't much care whether the body is charged or not. A magnet will have no effect on a charged balloon. But magnetic fields are created by moving charges and electric fields. So, if you moved at some speed relative to these currents, you could see the resulting magnetic field increase or decrease in strength. Your experience of these invisible electric and magnetic "dimensions" as an electron would be as puzzling as a human's experience of the space and time "dimensions" of the gravitational field. Einstein said in his description of gravity that time and space are really aspects of the gravitational field. As with electric and magnetic fields, themselves aspects of the electromagnetic field, time and space do not hang suspended and aloof from the bedrock of the gravitational field. Like the electric and magnetic fields that combine to synthesize what electrons experience as electromagnetic fields, time and space are the equally pliable components to space-time that we experience. This pliability can lead to some dramatic effects if you push them too far. According to special relativity, a place without gravity would also be a place without time, space, or the coordinates needed to describe this condition, just as an electric or magnetic field cannot exist independently of an electromagnetic field.

So where does that leave us?

We have added to the Void a set of ingredients called electromagnetic and gravitational fields, which act as nature's invisible partners to cause matter to move. Electromagnetic fields can be eliminated, but general relativity has now forced us to accept a startling conclusion: What we have casually called "empty space" is defined, created, and maintained by the gravitational fields of every scrap of matter and energy in the universe. Gravity is not a field described by only three dimensions of space. It requires a full four-dimensional description. Its intuitive properties of time, space, dimension, and shape are not features of some mythical Ether-like prior-geometry. There is nothing else "below" or beyond the gravitational field. Just as the electromagnetic field no longer needs the Ether to support it, gravitational fields no longer need some hidden Void or prior-geometry to give them their geometric properties, or even the attributes of time, space, and dimension. If you are looking for these reference points, you will have to look even deeper into the gravitational field itself to find them. It is in this great depth that we finally link up with the search for unity among the disparate fields and particles that clutter space-time.

As physicists tried to find unity among a diverse chorus of forces and particles, the search for an idea that could unify electromagnetism and the weak and strong forces had begun to meet its match. There was, at the very foundation of space-time and the physical world, something missing—some kind of principle or overarching idea—and its loss plunged calculation after calculation into a quagmire of impossible answers. Most of these answers involved the unpleasant word "infinity." Turning to even more abstract mathematics and tenuous analogies for guidance, one community of thinkers slogged ahead and tried to find the missing principle within the patterns they had habitually used. They did succeed in uncovering a dazzling new principle in Nature, or at least one that Nature ought to be using because it was deemed to be so beautiful and elegant: supersymmetry. Under some conditions, Nature might be able to convert the particles that transmit the forces into ordinary electrons and quarks, and back again. But if that were possible, then there must be a new option as well, and this one involved bringing gravity into the discussion through the back door. What physicists discovered as they explored the abstract terrain of supersymmetry mathematics was that when you converted one particle into another in the algebraic manipulations, the particles were forced by the mathematics to shift their location in space and time. This shift in space-time could only mean that gravity and a set of particles called gravitons carrying this force were intimately involved in the unification process. It was even more exciting that by including supersymmetry and gravitons, many of the problems physicists were having with their calculations disappeared.

By the late 1970s, armed with supersymmetry, physicists relentlessly pursued other remedies for new crops of problems, even with gravity and supersymmetry added to the mix. The most daring of these was to push beyond the four-dimensional description for space-time and consider an even bigger world with eleven dimensions. Physicists uncovered an incredible mathematical landscape. Our four-dimensional space-time, and the fabric of the gravitational field, was infinitely pockmarked by unfathomably tiny mini-universes wrapped into seven-dimensional spheres and other mathematical structures. We only experienced the four "big" dimensions predicted by Einstein's relativity, but subatomic particles were privy to a much larger universe. And in this universe there was the tantalizing prospect of finding unification. At the height of the enthusiasm for this new insight into particles and fields, there came a major disappointment. For the second time in a decade, physicists met with failure because the worlds they had crafted in their mathematics resembled nothing like our familiar world. There were different particle families, too many quarks, and a whole host of mysterious new particles that should have been long-since detected. Yet despite the frustration, a sort of rough

Plate 1: Reflection nebula in Orion. Darkness can be disturbing to us because it represents the unknown and unseeable. Shadows and dark places have always been disturbing to explore. In space, we often find odd contrasts between light and dark, and our minds still find some of them unsettling. (Courtesy of NASA/STScI; Hubble Heritage Collection)

Plate 2: Messier 83. Darkness is an important ingredient of the world, one that provides contrast and boundary to things that would otherwise be formless and devoid of detail. The dark lanes of dust that thread the arms of the Milky Way can trace their own patterns through the starry night sky and lead to sometimes terrible misunderstandings. (Courtesy of European Southern Observatory, Very Large Telescope)

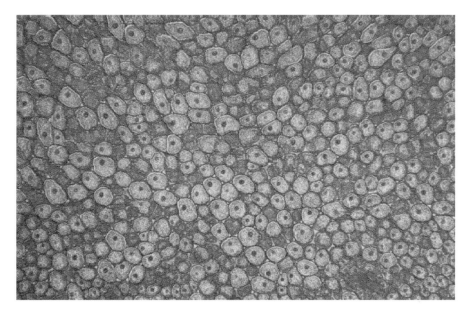

Plate 3: Nature has a fondness for bounding space with forms that have a cellular shape. Cells in living matter crowd each other while filling space as densely as possible. On the surface of the Sun, gases flow into patterns of convection cells nearly as large as Earth. (Courtesy of the author)

Plate 4: Quantum Corral. The atomic world operates by its own rules. This Scanning Tunneling Microscope image shows an artificial corral of atoms that imprisons an ordinary atom at the near focus of the ellipse, but a ghost atom with the same properties suddenly appears at the second, far focus. (Courtesy of IBM Research, Almaden Research Center)

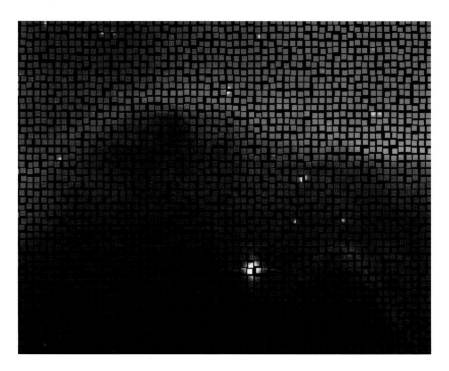

Plate 5: The vacuum that physicists study is the bedrock of all material things, and it is not a blank slate. It may be tessellated by domains of differing properties like this rendering of a distant nebula. (Courtesy of the author)

Plate 6: Scanning Tunneling Microscope photograph of silicon atoms. Only at the atomic scale do we see a slavish obedience to regularity and geometry. At larger scales, this geometric lawfulness is temporarily replaced by more fluid and organic patterns. The surface of a silicon crystal reveals the regular latticework of atoms. (Courtesy of Dr. Feenstra, Carnegie Mellon STM Group)

Plate 7: Galaxies by the millions form vast filamentary patterns spanning cosmic space in this computer simulation. (Courtesy of the Virgo Supercomputing Consortium)

Plate 8: Tomographic image of nickel foam. Space-time, as the embodiment of the gravitational field, may itself dissolve into a foam of interconnected elements or loops of energy that themselves have no smaller parts. This image gives a sense of the lacework structure that may inhabit the deepest levels of space. (Courtesy of Joachim Ohser)

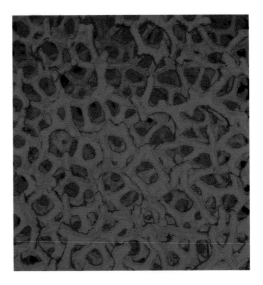

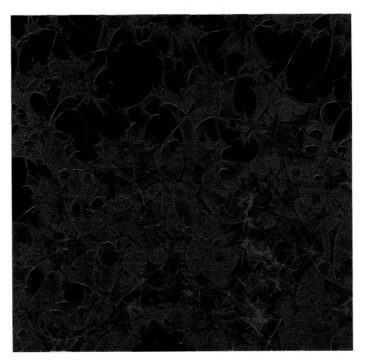

Plate 9: At its core, space may only be the tangled and shimmering confluence of waves of energy twinkling in the Void, but through quantum laws, it imparts space, matter, and field with the appearance of permanence. This computed image of the gravitational microlensing of light by a field of stars resembles the shifting patterns you see at the bottom of a swimming pool or the way that space-time fragments at its smallest scales. (Courtesy of Joachim Wambsganss, Potsdam University, http://www.astro.physik.uni-potsdam.de/~jkw)

Plate 10: Aurora photograph. Under the right conditions, invisible magnetic forces plying space can create visible curtains of light that dazzle the eye and fill us with a sense of awe and amazement. For thousands of years, auroras have graced the skies with their ethereal beauty. (Courtesy of Jan Curtis)

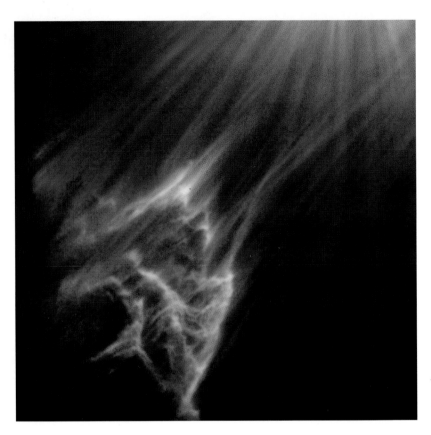

Plate 11: The Merope Nebula in the Pleiades. As we lose our sense of solidity and permanence, we discover that our bodies have more in common with the delicate, ephemeral substance of a distant nebula than with the rigid matter that fills our lives. (Courtesy of NASA/STScI, Hubble Telescope Heritage Collection)

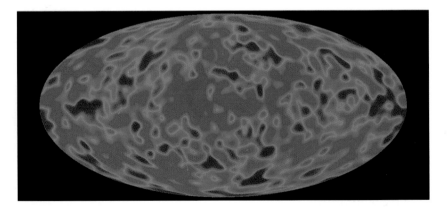

Plate 12: NASA/COBE satellite image. The ancient light from the Big Bang hides within its ripples clues to the origin and destiny of the universe. Sound waves frozen in time leave their marks in the subtle changes in this light across the sky. (Courtesy of NASA/COBE)

progress had been made. We now had a far less parochial outlook on space-time and the gravitational field. It could be a bigger arena than merely the four obvious dimensions it seemed to encompass, and even the concept of dimension had become unexpectedly malleable.

Could there really exist some kind of elemental Void, some unimaginable starting point for the physical world—that thing we may instinctively sense in the depths of darkness? Could this place or condition be the complete negation of every physical attribute we can comprehend? It would be a horrifically sterile place: timeless, space-less, and empty of content. Only the ephemeral gravity fields of our universe would provide a gossamer-thin veil separating us from this dimensionless abyss. In some unfathomable way, the gravitational field would contain all the information Nature needed to fashion time and space. Even the things it contained, such as the fundamental attributes of matter, would be part of its invisible fabric. Somewhere in this gravitational field, Nature has written the mandate that, for example, it can only be four-dimensional at its largest scales, that ex-actly three of these must be spacelike. But how is this information written within the invisible fabric of the gravitational field? In what language? How in the world does something as intangible as gravity prevent you from turning your ruler through a fourth right-angle to sample a fifth dimension to space-time? Could the answer be as sim-ple as the fact that we are here to ask these questions in the first place? Amazingly enough, we might have the means to answer even such an imponderable question with some logical assurance.

Einstein didn't intentionally design general relativity with these ques-tions in mind, but because it was wedded to the mathematics of Rie-mann, the logic of general relativity was more powerful than the four dimensions that tried to contain it. Like a telescope equipped with more than one eyepiece, general relativity lets us glimpse more distant worlds by simply dialing in a new dimensionality to space-time and seeing what happens. The results are spectacular. If space-time were three-di-mensional with only two space dimensions and one time dimension, gravity would not extend beyond the bodies producing it. In some sense, the entire webwork of space-time would dissolve into myriad un-connected points in the spaceless and timeless Void with no communication between them. If, however, we add one more dimension to our familiar world and make space-time five-dimensional, we create a new universe based on a gravitational field with four spatial directions and one time coordinate. Things start to look much more interesting. The spiritualists of the nineteenth century would have been very happy had our universe turned out this way. But there would be consequences for such a world that would be very annoying. You could, for example, never keep your shoelaces tied. You could enter and leave a room by seeming to pass through its solid walls. Then there would be a potentially fatal problem: Gravity could no longer be an "inverse-square" force but would morph into an "inverse-cube" law. Planets would not stay in stable orbits, thus stable orbits favorable to life would be rare. Only in a space-time with four dimensions do we seem to have the prospects for life-nurturing planetary orbits—and here we must marvel at its uniqueness. It's also amazing that Einstein didn't design relativity so that it would be capable of answering questions of this sort. These logical conclusions about other dimensions have emerged from the mathematical fabric of the theory of gravity, a theory specifically developed only to explain four-dimensional space-time.

So, the most important idea to carry away from this encounter with general relativity is that we do in fact live in the best of all possible worlds, with its maximum richness of potential detail. By some nearly magical symbiosis, it is our intimate and primordial relationship with the gravitational field, and the ineffable quality called dimension, that ensures we have this unique experience of a physical world. Exactly how this happens requires us to take an even closer look at gravity, well beyond anything that Einstein could have imagined. It would seem that there exist still other invisible confederates in the Void that busily go about the labor of knitting together gravity's web, preventing the world from collapsing into nonexistence. They also prevent us from turning a corner on the way to our local convenience store and stumbling into a fifth dimension. We must now search still deeper within gravity's field for clues to what these confederates look like and how they work their magic. Surprisingly, part of the clue we need is hidden not in its finest subatomic structure but in its breathtaking cosmic expanse.

6

ETERNITY'S ROAD
Cosmic Space and Its Expansion

Traveling eternity road
What will you find there?
—Ray Thomas, The Moody Blues

My college friend Dennis Reese and I were visiting his girlfriend, Gail Rogers, at her home in Oakland. After finishing a delicious meal, we sat around chatting about topics now long forgotten. As college students at U.C. Berkeley, there was no end to the many stories we had to share about courses, people, and events. It was twilight on a warm spring evening, and we had retired to the back porch to sip our wine and just hang out. I happened to look at the sky, with its typical urban cream-colored cast, when out of the corner of my eye, I spotted something moving. Our urban sky always had something going on in it, especially jets coming and going from the airports in Oakland and San Francisco. But this was different. All three of us stood and watched this curious apparition without uttering a sound. Gliding slowly from the north to the south at about the pace of the second hand on a watch was a cluster of lights arrayed in a V-shaped wedge, whose shape remained perfectly fixed. The distance between the eight equally spaced points of light never wavered as the formation silently sailed across the sky. The individual lights shone with a steady luster, never blinking. Given the size of the formation, about the extent of

113

your hand at arm's length, we fully expected to hear the din of the airplane engines these obvious "running lights" were attached to, but there was not a sound. I was sure this was no chevron of migrating birds in flight. Was it a low-flying jet? If so, where was the body of this plane? The episode was over within a few dozen seconds. We had all seen the same aerial event, but try as we could, there was no easy way to figure out what the thing was. As a future astronomer, I was frustrated about not being able to conjure up a sensible explanation. Even to this day, I haven't the slightest idea what we saw. By the strictest definition of the term, I had seen a UFO. For many people who are only casually familiar with the night sky, an even larger collection of phenomena can appear as UFOs, including the occasional fireball or passing satellite.

The sky is actually quite full of unusual things that come and go just below our limit of conscious perception. Easily seen events such as eclipses or comets prompt us to divine many different signs and portents from them. When you think about it, the sky is our most alien and hard-to-fathom vista. The starry dots can be connected into familiar patterns to give shape to a formless void, but it is nonetheless a very startling view, one the mind has difficulty accepting at times. The mind was trained to make sense of trees, rocks, and fleecy clouds, searching among these familiar landmarks for useful patterns to help it survive. The night sky provides none of this comforting familiarity. It is a black canvas flecked with luminous spots, which for much of our history as a species had no physical impact upon us.

Modern humans have always been fascinated by the sky and have worked very hard to paint a familiar face upon the unknown. It is unimaginable that the humans who survived the ice ages long before writing was developed didn't also have equally complex ideas about the starry firmament. What wonderful stories were told over campfires 30,000 years ago? Can you imagine, as I often do, what has been lost forever? Were there stories about the sky that would have put Shakespeare to shame? Thirty thousand years is a long time, after all, and there is very little that separates today's Homo sapiens from these ancient ancestors. There is one thing, however, that might have caused our ancestors to have a very different outlook on the world than we do now: our brains.

Our brains haven't always been the way they are now. The modern brain is split into two separate minds specialized to perform different tasks. The dominant hemisphere (usually the left side) holds the language centers and sees everything in terms of cause-and-effect patterns in time. It is the seat of logical-mathematical symbolism. It is also the place where consciousness resides as a jangle of background conversations with itself that you can overhear. The minor hemisphere (usually the right side) has no language centers and is therefore mute. It doesn't respond to cause and effect as much as to patterns in space. The problem is that it can't talk, so instead it passes powerful feelings into our conscious stream to make itself known. Every event is coded with an emotional response in the wink of an eye. In general, this is the place where various kinds of insights and creative connections seem to be forged. Some brain researchers speculate that the gradual advent of art, language, and symbolic cave paintings by our ancestors 20,000 to 40,000 years ago may have signaled the point at which our brains evolved into the specialized organs they are today. Our ancestors of 80,000 years ago may have experienced a more balanced worldview, one where the logical, cause-and-effect order of the left brain was not constantly overturning the holistically spatial, pattern-oriented ideas of the right brain.

Today, because of our brain structure, we have little choice but to see Nature as a series of events ordered logically in time. When you are lost in an unfamiliar city and then suddenly arrive at your destination, it is thanks to the work of your minor hemisphere and its pattern-recognition ability. Your dominant hemisphere usually says, "Aha!" as though literally seeing the answer contrived by the right hemisphere for the first time. Some of the greatest physicists admit that, like artists, they spend much of their time manipulating shapes and forms first, without mentally uttering a single word. Only after being guided by nonverbal hunches to find patterns that work do they create the logical and symbolic language to describe them. And sometimes the patterns physicists find can be so incongruous they are troubling.

On one of my camping trips into Yosemite National Park, I had spent a full day hiking a dusty trail and was anxious to set up camp in a shaded area. My map said that there was a brook nearby, and sure enough, a delightful stream of water flowed down a gentle slope under the shade of trees. I pitched my tent, thinking about the delightful dinner I would prepare in a few minutes, and as I worked I enjoyed the sound of the babbling brook. Little wonder that ancient tales from my native country, Sweden, were filled with wood nymphs and sprites who lived by brooks and streams. Later, as I sat on a rock resting my legs with the cool waters flowing across my weary feet, I could easily imagine hearing a word or two in the chaotic flow of the water over rock. But this daytime reverie took a turn for the worse once nightfall came and I retired into my tent for what I expected to be a good night's sleep. In the darkness, I could still hear the brook a few feet away, but now the random sounds knitted themselves together into a very different tapestry. Could I really be absolutely certain that *all* the sounds were from the brook, or had other campers moved into the site next to me? The babbling brook was confusing my ability to distinguish random sounds in Nature from perhaps more malevolent noises that would ordinarily have wakened me. Only a few miles away, a bear had taken one family's food the previous night. My mind worked overtime to try to separate the sounds and evaluate each of them. My right hemisphere kept finding patterns and dosing them with fear to bring them to my attention. My analytical left hemisphere dismissed them as implausible. In time, I did fall asleep, not from the physical weariness of a long hard day on the trail but from growing mental exhaustion.

When we look at the sky, it presents itself to our senses and our mind as a panorama beyond deciphering with any familiar logical system we possess, through either genetics or millennia of human experience. Trees and rocks are familiar, but not stars and the Void. To silently watch the night sky is to retreat into the nonverbal depths of the right hemisphere to search for some pattern or some meaning that will encompass the experience. The fact that none can, and the fact that the brain futilely searches in vain for some key to unlock the door of comprehension but finds none, causes us to stare with blank admiration and awe at the sky's vast panorama. Our minds are literally lost in space. Occasionally, however, we can personalize space and make some small part of it familiar—at least for a short while.

When my mother, Rosa, was a young girl around World War I, her mother, Josefina, died from tuberculosis. It was a horrible event for a seven-year-old to lose her mother. My mother happened to be at her aunt's farm in the country when the news came, and she became inconsolable. That night her aunt took her out into the garden and had her look up at the twinkling stars. A very bright one in the summertime firmament, perhaps Vega, caught her eye. She said, "Look, Rosa, there is your mother waving to you from her window in Heaven." My mother remembered that moment for the rest of her life, a moment when for a few minutes the cold, dark, starry sky became an instrument of personal solace.

Over the course of a human lifetime, there is something reassuring about the sky. Its most insignificant constellation looks the same year after year as our own lives play themselves out in time. All else might change for us, but the stars always look back at us with their unvarying patterns. What is the sky made of? What are all those lights? How far away is Heaven? For much of recorded history, people perceived the stars as scattered upon a distant canvas or sphere, far above the atmosphere. No one knew how far away they were. There is absolutely nothing about the sky that tells the brain just how far away the stars are, so thinking that they are only a few miles up is certainly fair game. There is also no penalty to be paid for guessing wrong. And it really doesn't matter to our survival as a species. No human action will be deflected by an inch, whether the stars are a hundred miles or a hundred million miles away.

It wasn't until 130 B.C. that the Greek astronomer Hipparchus, using some clever geometric arguments, decided that the Moon, Sun, and planets were actually rather remote. Back then, geometry was king. Even by Newton's time in the late 1600s, the size of the cosmos was still measured in millions of miles. It took another 130 years for the mathematician Friedrich Bessel (1784–1846) in Königsberg to figure out that some stars were actually trillions of miles away. By the early twentieth century, telescopic studies by Dutch astronomer Jacobus Kapteyn (1851–1922) made it quite clear that our universe was an enormous family of stars perhaps up to 10,000 light-years across, an unfathomable distance equal to 60,000 trillion miles. But was that all there was? Did the stars in the sky circumscribe everything that could possibly exist? Even by the turn of the twentieth century, there was no logical connection between the age of a universe and the distances to the stars. An eternal universe could be a thousand light-years across, or a billion. Was the universe small, or was it enormous and perhaps even infinite in size?

Harvard astronomer Harlow Shapley (1885–1972) and Lick Observatory's Heber Curtis (1872–1942) squared off in 1920 to publicly engage this debate, but the crux of the dispute was a question about the size of the Milky Way. Were the mysterious "spiral nebulae" located within the boundaries of the Milky Way as Shapley insisted, or were they distant external systems as Curtis believed? Curtis's evidence seemed the most persuasive, although both parties claimed to have won the debate. Yet even Curtis's scale for the universe, defined by the most distant spiral nebulae, was barely 3 million light-years—still an unfathomable distance to many at the time. Newspapers were excited about carrying this amazing news to the public.

It wasn't until 1929 that astronomer Edwin Hubble (1889–1953), working at the Mount Wilson Observatory in California, added forty more galaxies to this list, doubling Curtis's older estimate of cosmic dimensions to more than 6 million light-years. Eventually, Hubble uncovered a very strange effect that completely changed the course of astronomy: The farther away the galaxy was, the faster it seemed to be speeding away from us. Somehow, the speed of a galaxy was related to its distance from the Milky Way. This is an odd and completely counterintuitive result. For example, when you look outside your car window, you will see nearby objects moving faster than distant things, but this isn't what Hubble was seeing at all. The universe was doing something entirely different. Hubble's first estimate of the galaxy recession speed, the so-called Hubble constant, was around 500 kilometers per second for each million parsecs (mpc) of distance. Now an astronomer could simply use a spectroscope to measure the speed of a galaxy, divide the speed by Hubble's constant, and the galaxy's distance could be found. With this, the scale of the universe known to astronomers underwent a fantastic change.

By 1956, a catalog of the speeds of over 600 galaxies observed by astronomers at the Mount Wilson and Palomar Observatories included the fastest one then known—a galaxy that was moving at nearly 60,000 kilometers per second at a distance of over 1 billion light-years. This turned out to be only the threshold of a far deeper universe that sprang into view a few years later. In 1964, astronomers Alan Sandage (1926–) and Maartin Schmidt (1929–) discovered a perplexing class of starlike objects called "quasars," with speeds so enormous that astronomers were confronted with distances far beyond the 1-billion-light-year limit. The first one, identified simply by its catalog number, "3C48," was moving away from Earth at 37 percent the speed of light, which meant a distance of nearly 4 billion light-years if the Hubble law was still valid. The light now arriving at Earth started on its way through space at about the same time the solar system formed. By 2001, astronomers managed to refine their search techniques to uncover even more distant scraps of luminous matter at distances of over 10 billion light-years.

I remember hearing about quasars when I was in seventh grade. I was an avid reader of a weekly newspaper called the *Science News Letter*, which had just reported the discovery of the quasar 3C273 at a distance of over 2 billion light-years. But there was another nugget of information that completely dazzled me. It said that astronomers had used the brightness changes in this quasar to estimate a size of over one light-year for this "star." I was amazed that a star could be so big. I mentioned this fact to some of my friends and my homeroom teacher, Miss Zaik, the next day. She only looked at me with a kind of blank stare, which rather disappointed me. I learned the painful lesson then that most people rarely, if ever, show much interest in the stars or the universe. Even today, National Science Foundation surveys show that 40 percent of the general public doesn't even realize that Earth goes around the Sun once each year. So the distant quasars became my private little secret, whose exciting story of discovery I followed in the popular science magazines with the fervor of a stockbroker watching investments.

TABLE 6.1 Historical Estimate of Size of the Universe, 1600 B.C. to 2000 A.D.

Year	Author	Method	Size
1600 BC	Babylonian	Religion	Finite
350 BC	Aristotle	Philosophy	Finite
100 BC	Lucretius	Philosophy	Infinite
1500 AD	Ptolmaic	Celestial Sphere	A few million miles
1600 AD	Post-Kepler	Solar System scale	100 million miles
1710 AD	Newton	Gravity Dynamics	Infinite
1780 AD	Kant	Philosophy	Infinite
1838 AD	Bessel	Nearest star	8.6 LY
1901 AD	Kapteyn	Milky Way	30,000 LY
1915 AD	Einstein	Philosophy	Infinite
1916 AD	Shapley	Andromeda Galaxy	300,000 LY
1922 AD	Opik	Andromeda Galaxy	450,000 LY
1936 AD	Humason	Hydra galaxy cluster	1.1 billion LY
1965 AD	Sandage	Quasar, Ton 256	1.73 GLY (z=0.131)
1967 AD	Lynds	Quasar, PHL 1304	13.53 GLY (z=2.064)
1983 AD	Schmidt	Palomar BQS survey	13.9 GLY (z=2.2)
1989 AD	Schneider	Quasar survey	18.2 GLY (z=4.73)
2000 AD	Stern	Quasar, RDJ0301+0020	18.98 GLY (z=5.5)

Note: "1 LY" means one light year and "1 GLY" means one billion light years. Quasar redshifts (z) have been converted into comoving radial distances with a cosmological constant of $\Lambda=0.7$, a Hubble constant of H=70 km/sec/mpc, and using the calculator provided by UCLA astronomer Edward Wright (http://www.astro.ucla.edu/~wright/CosmoCalc.html). Comoving radial distance is the distance that a light ray travels across space, taking into account the expansion of space. According to Big Bang cosmology, our universe is infinite but it has a finite visible portion that we cannot see beyond.

In a mere 100 years, our conception of the size of the universe has expanded over 1 million times. Consequently, our minds have had to stretch to try to keep the entire picture in focus. This is not only a frustrating task, it is almost pointless to seriously undertake it. Even astronomers don't try to see the whole picture all at once. We can't simultaneously keep in focus both the space of our own solar system and the spaces that loom at the cosmic scales. It would be like trying to see the geography of North America while at the same time being aware of the level of detail of every square inch of its terrain. Why would you want to? Because eternity's road is incomprehensibly vast, it is nearly impossible to define with any single mental construct. Astronomers find it much easier to break the vast cosmos into three distinct arenas of experience, each one 100 million times bigger than the previous one. Like viewing paintings by Titian, van Gogh, and Seurat in an art museum, it is much easier to consider the artists individually as we wander among the various galleries in the cosmic museum trying to knit the whole thing together as best we can. Just as art museums partition art history into such formal periods as Renaissance, Romantic, and Modern, the cosmic galleries bear the names "interplanetary," "interstellar," and "intergalactic."

Interplanetary space, our cozy backyard in the universe, is the most familiar of the three. One day it may even become the volume of space that humans come to know firsthand through space travel. By earthly standards, though, it is a vista whose scope I can barely comprehend. If the Sun were the size of a basketball, our planetary domain out to distant Pluto would stretch about two miles across, a distance you can walk comfortably in a half hour. In the real world, it would take you nearly 100,000 years to make the same journey to Pluto by foot. Earth would orbit the basketball as a pea-sized dot about ninety feet from the Sun. Although our most distant space probes have penetrated some two miles from our Sun in the last fifteen years, the nearest star system, Alpha Centauri, is still over 9,000 miles away. Direct human activity in space defined by the limits of the Apollo program would be a sphere barely six inches across centered on our pea-sized Earth, crossed by a light beam in under three seconds. To say that after forty years, NASA has only begun to have humans operating in space is an understatement. It is also an especially frustrating realization for those of us weaned since childhood on stories of interstellar travel and galactic empires. Eventual trips to Mars will only expand this sphere by thirty-five feet in perhaps the next fifty years. Unmanned spacecraft continue to explore the far corners of interplanetary space, but we are centuries away from human outposts on planets other than Mars. If we could continue to double our sphere of planetary exploration every fifty years, by 2100 we would have colonies operating in the asteroid belt, by 2200 we would be at the orbit of Saturn, and by 2300 we would have reached Pluto.

As far as the greater depths of the solar system go, there is no compelling reason for humans to burden themselves with the expense and hazards of human travel beyond the asteroid belt. The Apollo Program provides a good historical example of colonization attitudes concerning the part of the solar system that is financially accessible to us. Unlike terrestrial exploration, once we journey to a new location, we tend not to go back there to establish a permanent colony. It is not too hard to imagine that in 200 years this same attitude will secure us only one outpost in the solar system, possibly on Mars. Twenty-second-century space life will certainly not be anything like the kind of vibrant multicolony civilization so many science fiction authors write about. Mars will attract human explorers once, perhaps twice, and they will collect a few hundred kilograms of rocks and fossils after a perilous 200-day stay on its surface. Then an endless political debate will ensue about why humans rather than robots should ever go back a second time. This is the pattern we have established for exploring the Moon, losing thirty years of colonization opportunity. Mars will be no different because politics and well-intentioned debate always seem to triumph over the imagination, at least in the short term.

Although we remain fixated on the various planetary bodies that accompany the Earth in its journey around the Sun, there is another dimension to the interplanetary vista that remains mostly hidden. At the largest scales, only the Sun and planets fill the interplanetary Void and capture our attention. Space appears empty and uninteresting. In fact, it is crisscrossed by more than 10,000 asteroids and comets and a carpet of microscopic dust grains. Some of this dust can be seen during twilight. A faint band of light stretches from the west along the path taken by the planets. This "zodiacal light" is the light from the Sun, scattered by the innumerable dust grains in space. At any moment, Earth is pelted by millions of pounds of this interplanetary trash. Once in a while, a large body crashes through the atmosphere, leaving a city-sized crater, or causes the extinction of vast portions of the biosphere. At a still smaller scale, each cubic inch hosts dozens of atoms passing through at millions of miles per hour from the solar wind, shown in Figure 6.1. Solar radiation bathes every cubic inch of space with millions of photons and neutrinos, which fan out into the colder depths of space beyond Pluto. Interplanetary space also exhibits a complex network of gravity fields ruled by the Sun, mottled by shifting domains of influence from planets, wayward asteroids, or

FIGURE 6.1 Interplanetary space, our familiar home in the universe, is awash with a dilute wind from the Sun, seen here in a satellite photo, together with innumerable dust grains and meteoritic particles. (Courtesy of SOHO/LASCO Consortium. SOHO is a project of international cooperation between ESA and NASA)

perhaps a few comets. No matter where you find yourself, you are always under the influence of our nearby planetary neighbors. The impact of these gravitational influences is not enough to affect humans, although they can literally turn the tide of terrestrial oceans.

For Earth, interplanetary space is certainly the most likely source of major disruptions in life, property, and civilization. Although remote, the occasional solar storm fills the Void with a hailstorm of invisible particles that can disable satellites and electrical power systems or compromise human health. Meteors that are several meters in size flash across our skies every few months as spectacular fireballs, while smaller ones collide once in a while with cars, houses, and even people. Every century or so, bodies the size of a large building explode in the atmosphere or near the ground, flattening hundreds of acres of forests. Every year, sensitive military detectors pick up the Hiroshima-like detonations of large bodies that have entered the atmosphere. It is only a matter of time before one of these undetectable bodies collides with a town or a city in our densely populated world. A projectile of this sort would not be seen until it crossed the orbit of the moon, so our fate would be sealed in a matter of minutes. On a much longer time scale, interplanetary space can choreograph even larger impacts between planets, comets, or large asteroidal fragments. Since the 1990s, hundreds of rocky bodies with orbits that intersect the Earth's have been discovered. The current list includes nearly 400 objects fifty yards across or larger whose orbits will carry them within a paltry few million miles or less of Earth. In time, these collisions could cause upheavals in our biosphere, snuffing out billions of human lives, either in the minutes or hours to follow or in the cloudy, soot-filled darkness that would last for months.

If interplanetary space defines our familiar backyard—the sphere of eventual human habitation—interstellar space is enormously more unfamiliar. This domain, defined by the dimensions of our own Milky Way galaxy, contains all of the stars in our particular corner of the universe, along with many billions more. Imagine two grains of sand two inches apart representing the Sun and the nearest star system, Alpha Centauri. In our interplanetary model, they would be two basketballs separated by 9,000 miles. In this scale model, you could walk to Alpha Centauri in half a year. To replicate the same jaunt in space would take you nearly a billion years. A model of the Milky Way would occupy a flattened disk of more than a trillion sand-grain stars in a lens-shaped volume two miles across and 100 feet thick. You could drive across this model galaxy in a few minutes, but light would take 100,000 years to traverse the same stretch in the real world. Even binoculars and small telescopes only allow us to see the stellar lampposts within a few hundred feet of our solar sand grain. Most of the Milky Way is completely unknown to us, and even our most powerful telescopes give us only a hint of its most distant shorelines.

Interstellar space is so vast that we have to think of it as being permanently off-limits to direct human exploration. It is easy to dream about interstellar voyages and alien empires, but the harsh technological and social reality is that we will never learn directly about distant sunrises on alien shores. Even our unmanned spacecraft, traveling at a million miles an hour, would take tens of thousands of years to travel to the nearest parts of interstellar space, even if we lived in the core of a star cluster like the one shown in Figure 6.2. We have no right to wrap interstellar space in the cloak of some romantic idea of space-faring futures that are out of our collective grasp, yet who among us who has avidly read science fiction stories can avoid doing so? If humans can barely manage to create complex machinery that will last a decade without failing—an automobile, for example—how will they ever be able to conquer the thousand-year gulfs that separate the stars? Although this kind of direct exploration is out of the question, we can stand on our shoreline here on Earth and look out across the waters to the distant starry islands using telescopes. Since the 1600s, astronomers have been able to observe interstellar space with steadily more powerful telescopes, and they have learned much from this 400 years of intense scrutiny. That isn't the same as actually traveling there, but it is the best anyone is ever likely to do, even in the most optimistic future imaginable, using technology rooted in the known physical laws. And between the distant stars we hope to someday reach, we see the interstellar Void unfurl its variegated cloak.

Scattered between the stars are dust grains and gas glowing dimly in the light of their feeble radio or infrared emissions. They form a tenuous interstellar medium with about one atom per cubic foot, along with a few hundred dust grains per cubic mile. This is a far cry from the density of the air we breathe or the much more crowded interplanetary spaces. A cubic inch of air on Earth contains about 10 million trillion atoms. A television picture tube contains a vacuum even more rarefied, at 100 million atoms per cubic inch. Specialized pumps can achieve ultralow conditions with 3,000 atoms per cubic inch, still crowded by interstellar standards, and in fact about the same density as the gases that make up the Great Nebula in Orion.

It took a long time for astronomers to figure out that interstellar space is not a perfect vacuum. Of course, the ancients considered the heavens to be filled by the Fifth Essence, from which the planets themselves were congealed. But it was not until much more recent times that anyone really suspected that space was anything other than

FIGURE 6.2
Interstellar space is filled with
stars and gas in a nearly continuous blanket of matter that at
light-year scales can seem as
dense as the air through which
we move. Here, in this photograph of the dense stellar core of
Messier 55, we see stars separated by less than 600 billion
miles. By comparison, the nearest
star to the Sun is 4.3 light-years
distant. Compare this view with
tungsten atoms in Figure 3.1.
(Courtesy of ESA/VLT)

a perfect vacuum. From Lick Observatory in California, the American astronomer Edward Barnard (1857–1923) photographed the dark gaps in the Milky Way—the same ones that the Incas had watched with foreboding—and by 1907, he was pretty well convinced that those dark regions were vast obscuring clouds, not just places where stars were missing. In 1926, the English physicist Sir Arthur Eddington (1882–1944) pulled all the evidence together and argued convincingly that the "interstellar medium" filled the spaces between the stars in the sky. The dark clouds that Barnard had so carefully photographed over the years were simply denser clumps in this galactic atmosphere. Once radio techniques were applied to mapping the heavens beginning in the 1940s, the feeble radiations from hydrogen atoms were detected everywhere. The entire dark void of the night sky glowed with the dim light of hydrogen gas. Then the Space Age arrived. New technologies were brought to bear on these studies. Once astronomers learned how to map the gas in three dimensions, it became very clear that the Sun's location was unusual. Observations in the ultraviolet portion of the spectrum by NASA's International Ultraviolet Explorer satellite began to turn up an interesting picture of our location in the local Milky Way.

Picture one atom in a cube two inches on a side. This is the pocket of low-density gas in which we live. In our two-mile-diameter Milky

Way galaxy model, our "local bubble" extends about forty feet across and contains millions of stars. Only a few thousand are bright enough to see in the night sky. Beyond this, a denser interstellar medium fills the rest of the Milky Way in a more complex pattern of pinwheel-like arms. It is a complex medium clumped with dense clouds shaped by stellar gravity and pushed by radiation fields into complex filaments and cloudlets. On a clear night, many of these clouds are silhouetted against the star fields of the Milky Way as dark blotches and rifts. The Incas knew these clouds of insubstantial dust and gas all too well.

No matter what scale you select, interstellar space remains a rich container, as familiar to astronomers as the planetary oceans are to modern sea captains. Astronomers have uncovered many of its secrets, both chemical and physical. It is cluttered with dust grains and dead comets, laced with high-speed cosmic ray particles zipping across the Void at nearly the speed of light. Despite its remoteness and its differences from our own realm, we can still accept interstellar space as an extension of our own world. Over thousands of years, human civilizations have made it the stuff of legends, characterizing it variously as welcoming or repelling, in the end absorbing it into the collective consciousness. In modern times, we have come to know that even some distant events in this vaster arena can make themselves felt here on Earth. Over millions of years, supernova have flared up, many within a few hundred light-years of Earth. These blasts of intense radiation and cosmic rays are muffled by Earth's blanketing atmosphere, but they flood interplanetary space with a hellacious tumult of particles and fields probably lethal to humans.

In the heart of Aquila the Celestial Eagle, a spinning pulsar known only by its catalog designation SGR 1900+14 rearranged its powerful magnetic field, hurling a blast of gamma rays and X-rays deep into space. It was an utterly trivial event: The pulsar's ultraprecise ticking clock hardly missed a beat. Halfway across the galaxy, 20,000 years later, this burst of radiation arrived at Earth on August 27, 1998, at 3:27 P.M., Eastern Time. Although greatly weakened by its journey through space and time, the pulse of energy disrupted instruments on satellites, wreaking havoc in Earth's atmosphere over the nighttime Pacific Ocean. The atmosphere caved inward by twenty-five kilometers as the pressure from the blast wave invisibly collided with Earth. The ionosphere flared up with new charged particles as though illuminated by the daytime sun. All of this was caused by an event that happened in interstellar space long before humans learned to write or build their first stone monuments.

Other events in interstellar space could impact Earth. A passing star could disrupt the distant cometary bodies in the Oort Cloud, sending a rainstorm of comets into the inner solar system, perhaps pummeling the Earth over the course of centuries or millennia. Even the invisible, fleecy clouds that inhabit the Local Bubble may pose a celestial problem for us in 10,000–50,000 years. As we pass through these clouds that are light-years thick, the pressure from the gas will drive the solar wind back into the inner solar system. The Sun might even dim ever so slightly from the invading dust, plunging Earth into an ice age.

Until the 1920s, interstellar space was the outer frontier of the universe. All that you could imagine about the universe was defined by the very stars you could count in the sky. Then, almost overnight, astronomers discovered that this spacescape was only the Middle Kingdom in a much larger tableau. Dim motes of light in spiral-shaped clouds soon revealed in their distances a deeper realm stretching into the unimaginable depths of intergalactic space. As astronomer Ormsby Mitchel of the Cincinnati Observatory noted in his 1868 book, *Planetary and Stellar Worlds:* "Worlds and systems, and schemes and clusters, and universes, rise in sublime perspective, fading away in the unfathomable regions of space, until even thought itself fails in its efforts to plunge across the gulf by which we are separated from these wonderful objects."

The familiar confines of interplanetary space, from the perspective of our solar system, shrink to a grain of sand in our model of the Milky Way galaxy. To make sense of the cosmic vista, we must collapse the Milky Way to the size of a teacup so that we can begin to comprehend the distances beyond it in the intergalactic Void. Our nearest galactic neighbor, Andromeda, itself a majestic twin to our own Milky Way, becomes just another teacup at the far end of our kitchen table. Nearly three dozen other teacups clutter the space inside our kitchen, which astronomers call the Local Group. If we step outside the kitchen door, we can see the houses in our neighborhood, which in our model become the homes of hundreds of other teacup clusters as we step back to view the Local Super Cluster of Galaxies, whose home office is the Virgo Cluster with its 1,000 galaxies. Then, at the outskirts of our city forty miles away, we encounter the farthest homes/teacups that can be seen in our visible universe. There may be other cities beyond, but they are too remote for light to travel the distance to us since the birth of the universe some 15 billion years ago.

In many ways, intergalactic space contains much more detail than any other type of space. Galaxies are large compared to their distances from one another, so our model is richly populated with easily recognizable shapes. The universe comes alive in a rich tapestry of galaxies that crowd out the dark gaps between them. Small binoculars are powerful enough to reveal the fuzzy glow from many of the teacups within our kitchen. Large amateur telescopes can discern many more teacups within our neighborhood. But only the largest telescopes, nearly as big as a small house, can see the light from the teacup homes at the outskirts of town and beyond, as shown in the galaxies in Figure 6.3. Astronomers find these remote outposts of galactic matter familiar and comprehensible, but it has taken a full seventy years to begin to map out our own neighborhood and what lies beyond. Even so, it is hard to fathom many of the details in the greater universe. The distant cosmos remains utterly mysterious because only its biggest metropolises have been studied and cataloged. The most magnificent of these have no familiar names, only obscure catalog numbers in astronomers' tables. As we observe these fantastic depths of space, we are also taught a valuable lesson about seeing the shape of the universe at cosmic scales: What we see is not really there.

We live in a phantom universe, a place where there is a lack of synchrony between where things appear to be located today and where they are actually located, due to the sheer size and age of the universe. Looking ever deeper into space, we get even more out of sync between events happening "now" and what the images carried by light tell us. The things we see paint a picture of a "now" that is as insubstantial

FIGURE 6.3 Intergalactic space is an arena bent by gravity and filled with islands of visible, luminous matter. Some of the galaxies in this view are 10 billion light-years from Earth. (Hubble Deep Field image courtesy of NASA/STScI)

and meaningless as a mirage on a hot summer day. We scarcely notice this problem for nearby planets and stars because they change little over the time it takes light to bring us information about them. Communication satellite operators and users find it annoying that there are light delays of a half-second in bouncing radio signals from place to place on Earth. Computer data streams can become badly out of step because of these delays, which are dictated by the laws of physics. The light from Jupiter makes a forty-five-minute journey to Earth, showing us where the planet was, not where its motion has now carried it in the meantime, a difference of about 22,000 miles along its orbit. Were it not for the fact that gravity also travels at the speed of light, the distant universe would dissolve into a bewildering chaos of hidden causes that would affect us without the comfort of our seeing them before they arrive. Gravitational forces depend on where a body is located at a particular time. Because the force travels at the speed of light, the only thing that makes a difference in the way a distant object affects us is where its light image tells us it is right now. It doesn't matter that by the time the light has traveled to the Earth, the star has moved to some other spot. That new location will only make a difference to a future astronomer when the light image and its new gravitational pull finally catch up to us. As we look into the distant universe, the space we see becomes more and more distorted because of this light-travel delay. Cosmological models can tell us what the universe really looks like today. By adding into the models the various light-travel delays and the curved trajectories that light must take, the models can also tell us how the universe actually looked from a particular vantage point at a particular moment in history. However, when the delays become very long, the universe ceases to be a physical system that telegraphs its punches. We don't have to go very far into space before we begin to see some of these interesting effects.

By the time images of galaxies 100 million light-years away can come into view, their forms have turned around on their axis by almost one-half a full rotation. By the time images of galaxies in distant systems 300 million light-years away reach us, a full Galactic Year has come and gone. Galaxies that we now see on the verge of collision at this distance have long since interpenetrated and taken up a new form entirely. In the far greater depths of the cosmos, time becomes even more severely unhinged. Because 300 million years spans the lifetime of a star capable of becoming a supernova, the empty cloud we now see in a remote galaxy could already have born a massive star that has run through its entire life cycle and has just now supernovaed. The light from this explosion will not arrive at Earth for another 300 million years. By the time we reach the distance of a few billion light-years, the galaxies we see in a tight cluster may already have shredded each other to pieces in an act of galactic cannibalism to form a monstrous snowball-shaped form. Some clusters such as Stephan's Quintet have already profoundly changed their shapes as powerful tidal forces rearrange their stars. At 10 billion light-years, the images of galaxies show them as they looked at the time they were being formed. Galaxies that are just forming no longer bear much resemblance to their adult form 10 billion years later, much as human fetuses don't resemble themselves as adults. The images of what distant galaxies look like now are still en route across intergalactic space. By the time they arrive 10 billion years hence, our own Sun will have been extinct for nearly 4 billion years, and there are unlikely to be human observers to record these images. Much of the way the distant universe actually appears today is forever lost behind the veil of the remote future. We see only the phantom images of how things once were. It is only the gravity from these phantom masses that steers our stars and buffets our home galaxy. Only the regularity of gravity and the dependability of the laws of Nature let us recreate the universe as it looks today from the God's-eye vantage point of an instantaneous, cosmic "now." But first we have to know where they are in the convoluted "now."

Since the 1940s, astronomers have continued the difficult task of mapping the cosmos literally one galaxy at a time. Using a variety of techniques that they compared meticulously one against the other, they collected the outlines of patterns in the cosmic Void in their tables. The first to be catalogued was the Local Group, a collection of thirty galaxies forming the Milky Way's family of neighbors. The next effort resulted in the deep catalogs of a thousand more distant clusters of galaxies, assembled by such astronomers as Fritz Zwicky (1898–1974), George Abell (1927–1983), and Gérard de Vaucouleurs (1918–1995). In time, the clusters of galaxies began to reveal still larger "superclusters" of galaxies containing dozens of clusters spread out like some great spider's web. If each of the billions of galaxies in our universe were replaced by a star, the vista you would see would be even more alien than the sparse stars that light up the Milky Way. You would see a carpet of light covering the sky, nearly completely filled in with individual galaxian points of light, the mottled outlines of arcs, and clumps of light. In these shapes outlined by millions of galaxies, patient astronomers can trace clues to patterns that lie beneath the seemingly implacable Void.

By the 1980s, vast "walls" and "voids" emerged from the galactic tapestry as surveys probed landmarks in space stretching 1 billion light-years from the Milky Way. Then in 2000, it seemed this uncovering of structure had ended. After studying the locations of 100,000 galaxies to a distance of 4 billion light-years, the astronomers Gavin Dalton from Oxford University, along with Karl Glazebrook from John Hopkins University, came up empty-handed. There was no evidence of any collections of matter in the depths of intergalactic space beyond the fantastic galactic ensembles 300 million light-years across. There was nothing larger than this.

Interplanetary space defines the region humans may eventually come to know through direct contact, and interstellar space is the domain of the untouchably remote but familiar night sky. Intergalactic space, the region beyond, must forever be the domain of our dreams and fantasies. Each teacup offers us a brew as richly complex as our own Milky Way, steeped in the perpetual mystery of the incomprehensibly alien. Nearby galaxies, under the probing eye of the twelve-ton Hubble Space Telescope, resolve themselves into beautiful patterns of millions of stars, each formed by the same processes as our own Sun, cloaked in eternal mystery. In our own island universe, the Milky Way, we can still hold out the hope of eventually discerning the feeble light from planets perhaps similar to those in our own solar system. However, we must eventually admit that even the finest technology we might imagine will never allow us to view the planets orbiting stars in other galaxies with good clarity.

Astronomers describe whole galaxies with a handful of numbers entered into a table. Many entries look nearly identical, but numbers hide the individuality of billions of stars and millions of worlds hidden in their foggy undetectability as securely as the quantum mist that hides a single electron moving in its own eternal darkness. When galaxies collide in a 100-million-year gravitational choreography, stars are stripped from their home galaxies, then tossed into the depths of intergalactic space. The infusion of fresh dust and gas into the galactic core regions feeds billion-sun black holes, which pour massive blasts of radiation deep into the disheveled galactic disks. Innocent worlds that managed to remain behind are blasted with radiation. Space travel becomes hazardous for millions of years. Some parts of a galaxy may be more harmful to life than others. The luck of the draw may decide whether promising worlds by the millions remain barren or explode with a profusion of life forms. When stellar orbits pass close to one another or close to vast interstellar clouds, gravity can dislodge a hailstorm of comets that will penetrate deep into a planetary system, bombarding orbiting worlds with devastating collisions. Our own Sun orbits the Milky Way far from the dense confusion of the core region, two-thirds of the way into the distant outer zones. As far as astronomers know today, the Sun was formed in just about the last possible region where interstellar clouds can even form and beget stars. With few neighboring stars to jostle the comets into deadly orbits, life can gain a foothold and begin its conquest of eternity.

If years or even centuries span the scale of interplanetary events and millennia encompass the interstellar episodes that connect with the human scale of existence, what then do we make of the billion-year epochs that bring the intergalactic domain into the fabric of human history? The only known episodes humans have yet to experience on these vast scales involve the eventual collision of the Andromeda Galaxy with the Milky Way, three billion years hence. Perhaps the most interesting impact of the cosmos upon humans is its legacy. Intergalactic space is not a messenger of future calamities so much as it is a marker of episodes that have already come and gone. Like historians or tea-leaf readers, we can read the intergalactic Void for clues to our origins. Even hints to the origin of our own universe are hidden there.

The entire gamut of cosmic space is now revealed for us to ponder, from the intimate confines of planetary realms to the incomprehensible vastness of the dark galactic Voids. If we want to explore space and the Void itself rather than the material contents that catch our eye on a dark night, the familiar landmarks of the things we can see directly have to be set aside. We have to look into the tide pools and beneath their seemingly calm surface to find clues to hidden mysteries and the great ocean beyond the shoreline. We have to avoid our own reflection in this still surface water and try to examine what remains, ignoring stellar or galaxian matter in all its fascinating diversity. Nothing in our genes or in our collective knowledge after millions of years of evolution can prepare us for what we may find in this cosmic abyss. Distant stars and galaxies are made from matter and controlled by gravity, the same way local matter at the interplanetary scale seems to be. How should we go about exploring the Void itself as a physical system? If the Void is space-time, and space-time is gravity, it makes sense to use the description Einstein provided in general relativity to guide us further. Nearly eighty years ago, the first attempt was made to describe the "big picture"—what the universe looked like and where it was all going—but it didn't lead to just one answer, it led to three.

Alexander Friedmann (1888–1975), a Russian mathematician, was the first to discover that there were actually three ways to describe our universe by using general relativity—three ways that space-time could wrap and distort itself to create the cosmic landscape that we see between the stars in the night sky. These were not universes that persisted indefinitely as motionless galaxies trapped in the Void. Friedmann's cosmologies demanded a dynamic, evolving universe in which the cosmos expanded or contracted under its own gravitational forces. Einstein had offered his own cosmological solutions nearly a decade earlier in 1915 and wasn't that impressed with Friedmann's conclusions. Einstein even offered a comment about Friedmann's work, saying that "Friedmann's paper, while mathematically correct, is of no physical significance." It's not that Einstein was a bad sport, but after dismissing Minkowski's geometry as superfluous learnedness, he made a second grave miscalculation about what his scientific competitors were trying to accomplish. Most likely, the idea of a motionless universe was too overpowering to ignore, even by someone as gifted as Einstein. Look outside your window at the night sky. You will see exactly what Einstein saw: Nothing moves. Intuition demands that there should be only one way to describe our motionless universe. In that way, it is eternal and unchanging. This was not, however, the way that the mathematical sea chose to part itself according to Friedmann. Also, when Einstein was crafting his own ideas about space, the universe was thought to be only as big as the Milky Way, a scant 10,000 light-years across. The distant galaxies were not yet considered much farther away than the stars themselves. Still, Einstein imagined that the carpet of stars extended infinitely in space. To prevent it from collapsing under its own weight, he had to add an antigravity force to keep the whole system eternally balanced. Friedmann's universes were very different. They rejected the need for this ad hoc antigravity force and were based on Einstein's original unblemished equations of cosmology.

All three of Friedmann's universes began from an unimaginable event in the remote past, later derisively called the "big bang" by the British cosmologist Fred Hoyle (1915–2001). Each showed a universe expanding and increasing in volume. One model even had enough matter in it that the combined gravity could actually stop the expansion altogether, reversing it into a collapse—the "big crunch." In many ways, the traditional view of Creation as a singular cosmic event in time had found its echo some 4,000 years later in Friedmann's mathematics. But there was a good deal more to these models than a simple rediscovery of the biblical moment of Creation. There was the very real prospect that the models would lead to prediction of the destiny of our universe, just as it is possible to predict the next solar eclipse or the flight of a baseball.

In the hands of Einstein, Friedmann, and others during the 1920s, cosmology became a full-fledged area of physical science, departing the realm of philosophy and religion. It was a far different way of describing the cosmos than anything previously offered within the scientific community. In nineteenth-century books on astronomy, when physics was ascendant across a wide spectrum of topics, there are countless descriptions of the enormity of sidereal space as revealed by astronomers. These works often contain speculations about distant island universes beyond the Milky Way at the limits of telescopic sight, hanging in the depths of space, filled with millions of stars. Space was perceived as a dark, passive container that served only to define the interval between galaxies, not to steer their courses. It was a dark abyss, punctuated only here and there by stellar light from galactic island universes shining weakly across unimaginably vast depths. To some popularizers, this enormity of space was a chilling prospect. The American popular science writer Garrett Servis, in his 1909 book, *Curiosities of the Sky,* offered this evocative description:

> Eternity of time and infinity of space are ideas that the intellect cannot fully grasp, but neither can it grasp the idea of a limitation to either time or space. The metaphysical conceptions of hypergeometry, or fourth-dimensional space, do not aid us. Having then discovered that the universe [Milky Way] is a thing contained in something infi-

nitely greater than itself . . . what conclusions are we to draw concerning the beyond? It seems as empty as a vacuum but is it really so? If it be, then our [Milky Way] is a single atom astray in the infinite; it is the only island in an ocean without shores; it is the one oasis in an illimitable desert. Then the Milky Way, with its wide-flung garland of stars, is afloat like a tiny smoke-wreath amid a horror of immeasurable vacancy . . . we are driven, then, to believe that the universal night which envelops us is not tenantless . . . and now the evidence assails their reason that what they had regarded as the universe is only one mote gleaming in the sunbeams of Infinity.

No one imagined that the entire cosmic system had a future that could be described by anything as concrete as a mathematical equation. As writers talked about the magnificence of these vast distances, no one thought of mentioning where the universe was "going." It was not considered an evolving system that was headed somewhere with a definite destiny. It simply existed. Only Newton had offered any new perspective on its fate. He found that its future was locked into the issue of whether it was finite or infinite. Only an infinite universe could be gravitationally balanced so that it would be long-lived and eternal. But it didn't take long before someone realized this kind of "cosmological solution" was as unstable as a pencil balanced on its point. Even Einstein's antigravity force could not save the universe from an equally devastating rush toward infinite dissolution. If all bodies had to be arrayed so that they canceled the gravitational pulls of all others, the whole material plenum would collapse just as soon as one of the points shifted slightly. Even the popularizers could see that there must be motion among the distant galaxies.

We do not know that gravitation acts beyond the visible universe [Milky Way], but it is reasonable to suppose that it does . . . Just as the stars about us are all in motion, so the starry systems beyond our sight may be in motion, and our system as a whole may be moving in concert with them. If this be so, then after interminable ages the aspect of the entire system of systems must change, its various members assuming new positions with respect to one another. (*Curiosities of the Sky,* pp. 14–15)

The Friedmann cosmologies now gave a new face to the problem of cosmic gravity. Since Friedmann's work was anchored in Einstein's general relativity, it had the best possible parentage. The solutions, however, automatically treated the matter in these distant galaxies and the cosmic gravitational field as one integrated system. Matter curved the space; space in turn controlled the motion of matter. Like a Gordian knot, it was impossible to sever the tie between them. The biggest casualty was the idea that space was immutable, with galaxies moving about like cinders ejected from a fireworks display. In Einstein's hands, space could no longer be a passive container. The 1919 solar eclipse had already proven that gravity could warp space. On the cosmic scale, this warping could become so severe that deflections would grow from a minuscule fraction of a degree around a single star to 360 degrees in the presence of the full material content of space. Space could literally close in upon itself, as all paths of light and matter looped endlessly around in a closed circle. There was, however, another aspect to the Friedmann models that confounded intuition completely.

All of Friedmann's models confound the mind more than almost any other idea in modern physical science—even infinity itself. They predict that three-dimensional space will grow larger as the universe grows older. Every one of us, astronomer or not, cannot help but visualize the expansion of the universe as some kind of titanic fireworks display that takes place in a greater, infinite darkness. The expanding embers are the outward-rushing galaxies, and we stand inside one of these and see an expanding cosmos all around us. It is a very comforting image created by our mind's eye. It has a rich parentage in the fireworks displays we have witnessed firsthand or on television. No matter who you are, you have the absolute conviction that space is just some kind of fixed stage within which actors move about without disturbing the scenery. But this is not what general relativity says is going on at all. Matter warps space, and space controls the motion of matter, in an intimate pas de deux. This means that the universe expands, but not by passively appropriating new volumes of space at its outermost edge where the fastest particles have arrived. The universe expands by streeeeeeeeeetching. This all takes place not within some larger arena of dimensional space but within an arena of pure nothingness. It's a bizarre idea, but one that scientists are absolutely compelled to consider, according to both the mathematics that describe cosmic gravity and the weight of astronomical data supporting it. Just as the 1919 eclipse required that both the gravity of the Sun and its warping of space had to be combined to give the right answer, the destiny of the universe is decided not just by the motion of matter but by how the invisible Void reacts to it.

Edwin Hubble's discovery that galaxies were speeding away from us in proportion to their distance was the next critical prediction by general relativity to be confirmed after the famous light-deflection test during the 1919 solar eclipse. This kind of motion, which astronomers call the "cosmological redshift," is exactly what general relativity based on space dilation had anticipated. It was not something that Newtonian models could even approximate. We can accept the idea that the universe may have had a beginning in time, and we can even accept that the universe may one day recollapse in a blaze of fire. But the mind will simply not accept that the expansion effect described in big bang cosmology isn't just the movement of galaxies away from each other in a fixed space like the glowing embers of a fireworks display. The mind has a much easier time accepting space as fixed and infinite than it does acknowledging the prospect that space, this empty Void between stars and galaxies, may be capable of stretching. Even as an astronomer, I have a terrible time working with this idea. It shows us just how badly Nature continues to confound us as we look outside the box of our limited world in space and time.

It is impossible to observe space dilation. We are much too small and our instruments are much too limited. That this happens at all can only be *inferred* from observations of the cosmological redshift, which general relativity then *tells us* means that the universe is expanding. Astronomers can intuitively imagine galaxies moving about like balls on a billiard table, but they have absolutely no conventional evidence that galaxies move. This may surprise you, but consider this: Astronomers can, over the course of years, show that the positions of nearby stars in the sky actually change. The shifts in the positions of distant galaxies over the course of a human lifetime, or even over a million years, is impossible to detect. Cosmological expansion is not a form of movement that any human has ever experienced. It isn't much of a surprise that our intuition reels at its implication, seeking other less radical interpretations for it.

All of our intuition is based on the idea, rooted in everyday events, that things move because they change position relative to fixed markers such as trees, houses, or mountains. General relativity tells us that this is not really the whole story. Motion as we experience it is simply the change in the distance between two markers. A bit of reflection shows that this can happen in two ways: Either the coordinates of the bodies can change in a stationary space, or the coordinates can remain the same but the space can dilate. The reason that cosmic expansion is not a motion through space is because, according to general relativity, the coordinates of the galaxies do not actually change. Instead, it is a motion *of* space. The coordinates are features of the gravitational field (space-time), and it is the field that dilates, carrying the coordinates with it. It would be like the latitude and longitude of Boston and Cairo staying the same while Earth expanded in size, increasing the distance between the cities without changing their location on the globe. The cities do not move on the globe, yet their separations increase. This is why some science popularizers like to use the expanding balloon or raisin-bread analogy when they describe the universe. It is not the raisins or spots on the balloon that move but the stretching of the dough or the rubber membrane between them that causes their separations to increase.

However, the analogy of the inflating balloon is not very accurate because it provides an unphysical vantage point for viewing the universe. Just where are you when you see the balloon's surface changing? If you are standing on any one spot on the balloon's surface, all other spots rush away from you as the balloon is inflated. There is no one center to the expansion on the surface of the balloon that is singled out as the center of the big bang. Even if you made this model an infinite rubber sheet, you would still get the same "centerless" answer. This is very different than the fireworks display, which does have a dramatic, common center to the expanding cloud of cinders. More important, the fireworks radiate from a fixed point in three-dimensional space that is the origin of the explosion. The balloon analogy, however, is not perfect because as we watch the balloon, our vantage point is still within a larger "something" that general relativity says never existed for the real universe.

The center of the big bang was not a point in space, it was a point in time. That makes it a well-defined center, not in the three-dimensional fabric of the balloon but outside it along the fourth dimension. We can't see this point anywhere if we look inside the space of our universe out toward the distant galaxies. But if we knew how to look, we could see its ghostly, ancient image written in the histories of every particle of matter in the universe. These histories all end, very thoroughly, at the big bang some 15 billion years back in time. Every scrap of matter, every clock in the cosmos, records its earliest moment starting from this common event.

In my opinion, balloons are good for parties, not for thinking cosmically. I prefer watermelons, the oval kind. Before we make a cut, the watermelon is a single entity. It exists as one single object embedded in the space of our backyard. The pulp of the watermelon, imprisoned by its skin, only exists within the volume of the watermelon, so that any equations that describe the dynamics of a "pulp" field begin and end inside the corpus of the watermelon; they do not apply to the nothingness of the "Void" outside it, and in which it may be embedded. Similarly, general relativity treats gravity as a field whose corpus is space-time. There is no gravity or space-time outside space-time itself. Whatever may lie beyond the confines of our space-time is not connected to our space-time. A hint of what may lie beyond our own space and time cannot even be seen in the symbols that fill our equations.

Let's imagine the cosmos of the sliced watermelon. We pick the longest dimension of the watermelon and slice circular disks perpendicular to this direction. Scanning the watermelon from one end to the other, we observe a sequence of wafers that increase in diameter until the midsection and then decrease in diameter to the other end. Now let's imagine this from the vantage point of a bacterium moving inside the watermelon along the long axis from end to end. As it passes through each slice, the bacterium sees its wafer world increasing in area until the area reaches some maximum value; then the area begins a decline until the point where the bacterium reaches the other end. The bacterium may well ask, "Where is the additional area of this wafer space coming from?" as it travels its path from one end of the watermelon to the other. Since the bacterium can only (by our construction) experience one slice at a time, this question of the changing area of its wafer universe seems entirely natural and proper. In fact, from our vantage point, the missing area of space the bacterium worries about as it sees its universe change has always existed in the three-dimensional body of the watermelon that we see. The explanation of why space seems to stretch is not rooted in the properties of space or in the substance of the watermelon. Instead, it is rooted in why the bacterium chose to move along the axis of the watermelon, in why it can't "remember" both its past and its future.

So let's add a dimension to the bacterium's universe and see where this takes us. As we look at the three-dimensional slices of intergalactic space, we see spatial volumes that increase in size from the moment of the big bang. Sitting inside one of these volumes, space itself seems to be expanding. We wonder where the new space came from. From the four-dimensional perspective of relativity, it has always been there. The three-dimensional slices that preoccupy us are part of a complete four-dimensional object, the cosmic gravitational field (the pulp of the cosmic watermelon) that general relativity says exists all at once and is essentially timeless. Our predicament in worrying about where space comes from is analogous to the bacterium's asking where the extra pulp comes from as the slices in its wafer universe get bigger. We are not getting something—namely, space—out of nothing by some magical hocus pocus. Instead, the expansion of space is our introduction to an aspect of Nature in the fourth dimension that is beyond what evolution in a three-dimensional world, wrapped around the outside of a fixed planet, has equipped our common sense and intuitions to handle. Space isn't going anywhere; it has already arrived. Our biggest challenge is to understand the entire picture all at once and to think about the beginning and ending of time itself. Only then will we truly understand what space-time means as a physical concept. Only then can we truly understand gravity.

7

ETERNITY'S END
Dark Energy and Cosmic Acceleration

In the time when Dendid created all things,
He created the Sun;
And the Sun is born, and dies, and comes again.
He created the Moon;
And the Moon is born, and dies, and comes again.
He created the stars;
And the stars are born, and die, and come again.
He created man;
And man is born, and dies, and never comes again.

—African song

Nothing could be more exciting to a young would-be astronomer than setting off with a group of like-minded friends for a weekend in the mountains with our home-built telescopes. On this particular occasion, Rich Simpson, his brother Stan, and I were making our monthly trip to Mount Diablo just outside of Walnut Creek, California. The campsites near the summit were the favorite attraction for many amateur astronomers in the San Francisco East Bay region during the early 1970s, but this time we were to have the whole campsite to ourselves. As usual, we made our excursion in Rich's VW van, passing Walnut Creek and Danville, with a quick stop at the Alpha Beta market to pick up what teenagers claim is food. The winding two-lane road to the summit brought the California shrub oaks

143

within easy reach and the fragrance of dried California grasses soon filled the van. We arrived around 3:00 P.M. and took afternoon hikes on familiar paths filled with sticky brambles that covered our socks. As sunset fell, the stars began to appear in the sky, so we busied ourselves with setting up our telescopes for an evening of sky watching. Within a few moments, we were ready to explore our favorite celestial objects in the Milky Way, from Sagittarius to Cygnus and beyond. We drank Cokes from the cooler, devoured chunks of pastrami cut with a dull pocketknife, and scanned the heavens from one cosmic spectacle to another. The chatting was incessant, but often not about the things we were seeing through the eyepiece. There was always so much to say about the endlessly eventful high-school lives we shared.

As the Milky Way turned to the west, we said our good-byes to the ancient light of Andromeda and secured our equipment against the morning dew, but because of the considerable amount of caffeine we had consumed, we weren't tired. So we brought out the ten-band shortwave radio and settled down to surfing another, more invisible, dimension in our lives. For me, it was just incredible to be gazing at the stars while scanning the radio dial from end to end. I heard the sounds of programs from around the world in often unfamiliar languages. Interspersed with the human stations were sounds of a thoroughly mysterious quality. Rhythmic throbbings of unseen machines added their alien voices to the human din. Stations drifted in and out as ionospheric conditions ebbed and flowed. Loud and soft stations suddenly became closer or more distant, giving the radio noises a startling third dimension of depth.

In an utterly delightful synthesis, the stars in the sky merged with the richness of the electromagnetic window provided by the radio. I imagined that I was tuning into interstellar messages between distant civilizations. The Old Ones in the galactic core were continuing their debate on quantum philosophy. Elsewhere, a civilization long since vanished after its sun had novaed was still transmitting its entire history from an automatic relay orbiting Procyon. In the depths of space, I could also hear the faint electromagnetic ocean of the cosmic background, which resembled my own tinnitus. I could discern the shadows of galaxies formed a billion years after the big bang. And in the reverie of the moment, we sat wondering about our futures. What twists and turns were now in store for us? As the sky turned toward the dawn, we knew the same summer stars would always be there to welcome us no matter what our destiny, but out of all the possible destinies that seemed so real to us as teenagers, which one would be ours?

In the grander scheme of things, the complex equations of Einstein's general relativity also contained endless possibilities for worlds that might have been and destinies still to be reached. Yet only a few of these mathematical worlds are reflections of our own universe as it is, and as it will be. Like tuning a radio dial in search of distant stations, we have to scan through universes without number, drifting in the mathematical void of things that could have been, until like a beacon in the Void, we at last find the universe called home, with its familiar stars, galaxies, and planets.

Suppose you were sitting at a baseball game watching the Boston Red Sox. It is the bottom of the ninth inning. The teams are battling to win the World Series. The game score is 2–1, with the Red Sox trailing. The Red Sox are up to bat, bases loaded, two strikes. The pitcher releases a fastball and the Red Sox batter makes contact with it. The destiny of this entire baseball series now depends on which family of curves the ball will travel along after being hit by the bat. It mostly comes down to two numbers: the speed of the ball and the direction of the ball at the moment of impact. There are an infinite number of paths that connect the batter's location with myriad points in the baseball field. One set of possible paths will lead to victory; the other paths, to the mitt of some player on the opposing team. In less than five seconds, the countless possibilities resolve themselves into one destiny. The ball travels its path, this time speared by an outfielder. No victory today, but perhaps in some other universe just around the corner from ours, an alternate game plays itself out differently to an audience of ghostly spectators.

For our universe with its battle between speed and gravity, there are also two numbers that let us glimpse its destiny. In this case, the bat and the ball made contact 15 billion years ago. If we knew how to look, we could see the outcome of this game, already written all around us. Just as the outcome of the baseball game could have been discovered by measuring the direction and speed of the baseball, astronomers can tell whether the cosmic ball game will end early or go to extra innings. All they need to do is measure how fast the universe is expanding. This is balanced in the equations against how much restraint gravity can provide the outrushing galaxies. Astronomers call the first of these numbers the Hubble constant; the second is called Omega. Once they know how fast the universe is expanding today from the size of the Hubble constant, astronomers can decide how much gravity the universe needs to prevent the expansion from continuing without end. This critical amount of gravity reflects how much gravitating "stuff" is out there in the universe. Like a line of challenge drawn in the sand, it reveals when a universe is destined for eventual eternal expansion or future collapse. Once this critical mass is known, the stars, galaxies, and other forms of mass are counted to see whether it all adds up.

As we enter the twenty-first century, the best estimate for the Hubble constant seems to be near 70 kilometers per second per megaparsec, an unusual set of units for a physical constant, but no more so than "degrees" for temperature or "miles per hour" for speed. Astronomer Wendy Freeman and her colleagues have used the Hubble Space Telescope for nearly ten years to make high-precision measurements of galaxy distances and speeds for dozens of galaxies. They uncovered a very narrow range of values around 70 km/sec/mpc as their final call. This means that galaxies farther away from us are being carried by the stretching of space to even greater distances at a rate of 70 km/sec for every 3.26 million light-years (a megaparsec). A galaxy located 10 megaparsecs away (32.6 million light-years) is receding at 700 km/sec. A galaxy at 100 megaparsecs (326 million light-years) is receding at 7,000 km/sec.

The first thing astronomers have learned by determining the Hubble constant is that the universe is expanding, not contracting. This places the history of the universe within only a small number of possible types out of an infinite number of possibilities. The bat and ball connected billions of years ago. The ball is definitely headed into the

TABLE 7.1 Determinations for the Age of the Universe, 1500 B.C. to 2001 A.D.

Year	Author	Method	Age in Years
1500 BC	Brahmin	Religion	8.5 billion (2 Khalpas)
350 BC	Aristotle	Philosophy	Eternal
100 BC	Lucretius	Philosophy	Eternal
882 AD	Mayan	Religion	5125 (One Great Cycle)
1650 AD	Ussher	Biblical	5650
1760 AD	Buffon	Earth cooling time	75,000
1831 AD	Lyell	Fossils and orogeny	240 million
1897 AD	Thomson	Earth cooling time	400 million
1905 AD	Rutherford	Radioactive decay	1.6 billion
1915 AD	Einstein	Philosophy	Eternal
1927 AD	Lemaitre	Expansion Age	1 billion (H = 627)
1929 AD	Hubble	Expansion Age	1.4 billion (H = 465)
1940 AD	Gamov	Earth Age	6 billion
1952 AD	Bok	Galactic clusters	10 billion or less
1958 AD	Sandage	Galaxy recession	8.6 billion (H = 75)
1970 AD	Sandage	Oldest clusters	17 billion or less
1991 AD	Cowan	Element dating	13 to 21 billion
1994 AD	Sandage	Globular clusters	15 billion
1994 AD	Birkinshaw	S-Z Effect	12 billion (H = 55)
1995 AD	Whitmore	M87 clusters	8 billion (H=78)
1996 AD	Chamchan	Milky Way disk	12 billion
1997 AD	Fischer	Gravitational lenses	8 billion
1997 AD	Perlmutter	High-z supernova	13.8 billion (H=70, Λ=0.7)
1999 AD	Lineweaver	Statistical fits	13.5 billion (H = 68 Λ=0.65)
2001 AD	Beers	Uranium in old star	12.5 billion or more

Note: The symbol "H" is the value of the Hubble constant in units of km/sec/mpc. The symbol "Λ" is the cosmological constant. Many of the estimates subsequent to 1991 have uncertainties of about 1 billion years or slightly more. Only the average value is listed.

ballpark's outfield. It is not a pop fly or a line drive. These mathematical universes also tell astronomers how to use this number to estimate an age for the universe since the big bang. There are actually several ways to run the movie of our universe backward in time. But it is important to choose the right one to start with, or the answer will be misleading. If the universe has been expanding at exactly the rate suggested by 70 km/sec/mpc since the big bang, neither slowing down nor speeding up, the time since the big bang would be about 14 billion years. This estimate does not reflect the physical fact that gravity could have been slowing down the expansion all this time. Specific big bang models account for this, leading to an actual age for the universe that makes it 8 to 11 billion years old, which is a lot younger than some of the oldest stars we study in the Milky Way. Getting the cosmic age right is a critical test of theory if we are ever to accept what general relativity tells us about space and the Void. Surprisingly, the destiny of the universe comes down to what astronomers can discern from only a handful of ancient stars.

It was an old star system in a young universe. Over the eons, the unlikely pair of stars had reached a stable give-and-take in their dance through space. The larger of the two was a distended red giant, an obese star whose blood-red gases wafted out into space. Some of this gas was captured by its companion star, a small white dwarf no bigger than Earth, the only remains of an ancient star. The white dwarf corpse was once the core of a star in which powerful nuclear fires converted matter into light and heat, but now the fires had long since vanished from the scene, leaving behind a hot, inert mass a thousand times as dense as lead. As the hydrogen-rich gases from its red giant companion piled up on the white dwarf's surface, every so often they erupted into nova-like brilliance. Like gasoline thrown onto a fire, the blast of energy could be seen clear across the galaxy as the nova outshone the combined light from millions of stars at least for a few weeks, then faded away just as suddenly. The detonation was powerful, but not powerful enough to eject all of the new material that had landed on the dwarf's surface. Over time, through thousands of nova eruptions like some cosmic beating heart, the supply of surface gases continued to increase the mass of the white dwarf. Then, like a table collapsing from its accumulated load of books, a gravitational precipice was reached. The sheer weight of the dwarf caused the core to implode, triggering an incredible release of energy. The carbon-rich matter in the core was once again engulfed in a thermonuclear inferno. After millions of years, nuclear fires again raged in the core of this dead star. Carbon atoms fused with carbon atoms to produce nickel and iron atoms. Within minutes, the entire core of the dwarf was converted to iron. Only the surface of the dwarf escaped this detonation, but its destiny was only moments from being consummated. The fusion wave tore the dwarf apart, ejecting the star's matter into space. For months, the expanding cloud plunged into space at speeds of 100 million miles per hour. The radioactive nickel atoms in the cloud decayed to cobalt. Over time, the brightness of the supernova reached a maximum, then faded steadily. At its peak, which lasted a mere week or less, the supernova was brighter than 100 billion suns.

Eighteen billion light-years away, and 10 billion years into the future, a feeble flash of light from an indistinct galaxy in the constellation Pegasus passed across the Milky Way. On October 15, 1998, the powerful light-gathering instruments of the Keck II telescope in Hawaii greedily drank in the photons of light from this supernova, tracking its waning brilliance. The telescope just happened to be pointed in this direction of the cosmos at exactly the right moment to capture the event. Other instruments probed the influx of light energy, sorting the photons into frequency order to measure its spectrum. A quick calculation told astronomer Saul Perlmutter and his team that it was the most remote supernova ever detected. The rise and fall of the light lasted only a few weeks, more than enough time for Perlmutter to classify it as a "Type 1A supernova," one of the best standard candles for cosmology known to astronomers. They named it Supernova Albinoni out of a sense of the singularity of the moment, avoiding the formal designation "1998eq" meted out by the International Astronomical Union. Somehow, out of reverence for what the star was telling them and Perlmutter's interest in classical music, it deserved something better than a terse catalog number.

Supernovas are the titanic explosions of stars that unleash powerful bursts of light seen halfway across the universe. Astronomers can use this light to measure the distance to their host galaxies. From this they calculate the ancient values of the Hubble constant as it was billions of years ago. In our local neighborhood of the universe, the light from thousands of galaxies as far away as several hundred million light-years from the Milky Way enters the astronomer's telescope. This provides a snapshot of the local history of the universe as it was up to several hundred million years ago. During this time, galaxies have continued to move away from us at about the 70 km/sec/mpc determined by the Hubble Space Telescope survey. Astronomers can also measure the value of the Hubble constant in collections of still more distant galaxies that are over 10 billion light-years away by using supernovas. These galaxies reveal what the universe was doing billions of years ago, a still more ancient snapshot of the cosmos beyond our neighborhood. This is where the story becomes intensely interesting.

For the second time in two years, Perlmutter's team had found a supernova gleaming in the dark pool of the distant universe. Together, the two events seemed to be telling the same story about our universe. Both of the distant supernovas were measurably fainter than they should have been if the expansion of the universe had been slowing down the way the most favored big bang models predict. Supernova Albinoni is nearly 8 billion light-years farther away than the standard big bang models called for if the universe had been steadily slowing down its expansion all these years. This is similar to a friend telling you that he would drive no faster than 70 miles per hour when he set out on a 1,000-mile journey. You calculate how many hours it should take him to make the trip, then measure his speed with a radar gun and confirm that he is moving at that speed. But when he arrives at his destination, you discover that paradoxically, he got there in a lot less time. In the case of the supernova, when all of the possibilities were eliminated for measurement error, only one conclusion remained: The universe is expanding faster today than it was 10 billion years ago when Albinoni's light first started on its way. This means there is an even more powerful force at work than gravity that is steering the destiny of matter and propelling the expanding universe from its infancy into its current middle age. What could it be? The big bang models that account for accelerated expansion provide the only clue to the puzzle. Like a chameleon, the mysterious ingredient that causes accelerated expansion has gone by many names over the last eighty years: cosmological constant (1920s), vacuum energy (1970s), and dark energy (1990s). It is the oldest ingredient ever to have been introduced into the equations describing the cosmos, introduced by Einstein himself in 1915.

For decades, astronomers looked at the cosmological constant with suspicion because it seemed to be just an artificial way of forcing the cosmos to behave the way humans wanted it to for philosophical reasons alone. Remember, in 1915, astronomers didn't know that the universe was bigger than the Milky Way galaxy. Einstein, looking out his window at night, saw no celestial motion, so he doctored his equations to create a static and unchanging cosmos. The cosmological constant was the "fudge factor" he had to add to make his cosmic models agree with the commonsense expectation of a static universe. Einstein officially gave up his own idea for it in a famous "biggest blunder" confession in 1931, once Edwin Hubble's observations of an obviously expanding universe took center stage in cosmology. But there was no good evidence that forced cosmologists to throw this factor out once and for all, so there it sat. Out of logical necessity, astronomers were always forced to interpret their data as supporting two kinds of big bang cosmology: one with the cosmological constant, antigravity force; and one without it. Many studies spanning the last half of the twentieth century tried to nail down its value. Only a range of possibilities that often included zero came up as a statistical answer. By the 1990s, however, this all began to change as new instruments let astronomers probe the properties of still more distant galaxies. Supernova Albinoni was just the vanguard in a whole new generation of high-precision studies of the deep cosmos. Soon, a new ingredient called dark matter began to show itself in clusters, like EMSS 1358+6245 in Figure 7.1.

It wasn't just astronomers who puzzled over the cosmological constant. As physicists began to explore various prospects for a grand unified theory in the 1970s, they independently uncovered in the pertinent mathematics a new set of fields in nature: the Higgs fields. We encountered these mysterious fields in Chapter 5, and they all have the peculiar property of being an omnipresent but hidden ingredient of the Void. Physicists quickly realized that these new fields, or their numerous cousins, behaved in exactly the same way as Einstein's antigravity force. As they interact with each other in the Void, they cause a feeble pressure that exerts its influence opposite to gravity. Cosmologists had at last made contact with physicists, confirming that some kind of enormous new field permeated the Void. The good news for everyone was that it allowed the universe to be much older than original estimates had suggested. Thus, the universe was once again just a little bit older than the oldest known stars.

FIGURE 7.1   The cosmos is controlled not by visible stars but by vast reservoirs of dark matter and mysterious energies that remain unseen. These ingredients are one of the greatest mysteries of twenty-first-century astronomy. The cluster of galaxies is known only by catalog name EMSS 1358+6245 and is about 4 billion light-years away in the constellation Draco. This Chandra satellite image of the X-ray light from million-degree gas within this cluster allows astronomers to determine that the mass of the dark matter in the cluster is about four times that of normal matter seen in the dozens of galaxies that fill this same volume of space. (Courtesy of NASA/MIT/J, http://www.msfc.nasa.gov/NEWSROOM/news/photos/2001/photos01–298.htm)

Now it was possible that this dreadful contradiction, which had been sighted and resighted throughout the 1980s and 1990s, might be brought to a final resolution, but only if continuing studies reaffirmed that the universe was indeed accelerating its expansion as it grew older. Like the hints in 2000 that the LEP machine at CERN had detected the Higgs particle, the key observations in 1998–2000 hinged upon only one or two very distant supernovas—an awfully tenuous thread upon which to hang an entire cosmological revolution. By the time Perlmutter and his team arrived on the scene, the theoretical study was already a mature science. Physicists had long since looked into exactly what would happen if space were laced with anything resembling a Higgs field. By 2000, their version of the big bang, which they called "inflationary cosmology," was an idea nearly twenty years old. It had quietly weathered its own share of experimental challenges and confirmations. However, its greatest impact was not on today's universe but on an unimaginably brief moment in its infancy.

The first physicists to look into this in some detail were Alan Guth at MIT and André Linde at Moscow University in the early 1980s. In the cosmic equations that included these fields, they uncovered a spectacular change in the very early history of such universes that is predicted by big bang cosmology. For a very brief moment after the big bang, the universe would expand at an accelerated rate, doubling its size every one-trillion-trillion trillionths of a second. The phase ended after perhaps a few thousand doublings, by which time the infant universe had inflated from a size that was many trillions of times smaller than a proton to something as large as a basketball. All of the universe we see today could have emerged from a patch of primordial space-time smaller than a proton. By the time it was barely one second old, our cosmos would already have been billions of times bigger than conventional big bang cosmology predicts.

Only by blending the detailed physics of ordinary matter and field with the completely featureless physics of the gravitational field did this new possibility for cosmic evolution emerge from the mathematics. Physicists were delighted that they had finally addressed cosmological issues by starting from subatomic laws. The cosmological constant might be a leftover remnant from this ancient era, or it might be caused by the workings of an entirely new set of fields in nature that only now have begun to gain the upper hand. In the coming decades, astronomers expect to perfect their measurements of how strong this mysterious ingredient is, but knowledge of any of its detailed properties seems beyond reach, given the incomprehensibly vast scales it occupies in both time and space. The only hope of studying it seems to lie in recreating, via computer models, what the universe ought to look like today according to various assumptions about how this field works. It is possible that some trace of this new field can be dimly seen in the pattern imprinted on the filamentary webwork of mil-

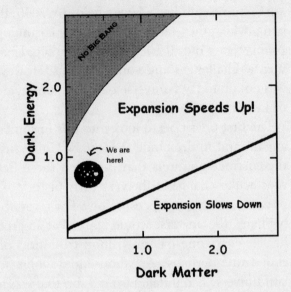

FIGURE 7.2
The destiny of our universe is circumscribed by the actions of dark matter and energy, which will propel it to an inevitable ending in the not-too-distant future.

lions of galaxies sprinkled through space. If so, we will know with a measure of scientific certainty just what kind of universe we have been born into, along with its eventual destiny. Given where the data seen in Figure 7.2 stand today, after fifty-plus years of hard work by thousands of astronomers, there is a good chance that we may not like the picture of the universe that is beginning to take form.

The empty cosmic Void expands, dark matter pulls, and hidden energies lurking in the ubiquity of space propel the universe to accelerating dissolution. On the nuclear scale and beyond, gravity's quantum turbulence shakes the Void. It is, in either extreme, the same field acting across time and space to define itself according to its own mysterious substance. Like a commandment echoing in the abyss, it brings itself into existence in the ultimate act of self-creation, not just once but billions of times a second. Its four-dimensional character makes it both the quantum "Alpha" and the cosmic "Omega," endlessly creating itself while at the same time remaining inert. Our own human experiences flourish in the middle ground of this process, and we are only dimly aware of the incredible richness, the profound subtlety of our existence, suspended as we are from moment to moment, a cosmic heartbeat away from nonexistence. Hidden activity works its invisible but relentless magic, knitting the gravitational field into a seamless plenum—its outer boundary lost beyond the farthest galaxies, its nearest limit lost in an equally remote quantum fog. With hard work, we can uncover glimmerings of the underworld in the canopy of activity around us, though we can never be sure that what we see in these unimaginably remote landscapes isn't mostly a figment of our imagination, projected upon the dark canvas of the Void. After all, humans have the innate gift of being able to see patterns, to divine destinies, from the merest scraps of information. The Void has always been a willing canvas, even for our darkest fears, but if cosmology has in any way touched upon the reality of our destiny, even ancient Inca fears seem tame by comparison. What cosmology has now added to the story of our existence has deep roots that extend not just into the dim ancient times of the infant universe but into the equally unseeable future, where we catch the first glimpse of what nature ultimately has in store for life in this cosmos. It is, however, not a story of continued evolution and optimism. The story we now see unraveling in the invisible machinations of the Void is the story of the final death. The Incas knew of one death; we now know of three: the death of the individual, the death of the living biosphere, the death of the cosmos. The last of these is controlled not by the things that are easily seen but by things that are as dark as the blotches in the Inca sky. It seems difficult to grasp the finality of cosmic death, but even in our own human sphere of experience, we can begin to understand what horrific implications it could have for our sense of perspective.

Within our genes and the strand of DNA that creates them, there is a silent clock beating out a cadence. Each time a cell replicates, the teleomeres that control the choreography of cell division get just a bit more worn. Finally, after a dozen or so generations, crucial information is lost; the cell dies. For the aggregate system that is the human body, there comes a moment when we accumulate enough of these little deaths to add up to our own mortality after fourscore years or so. If we are fortunate and if we have had a careful and lucky life, we can live to a ripe old age, limited only by our genetic heritage. Some of us even live past the century mark and into the uncharted human territory of 120-plus years. But then we die. With each of us goes a lifetime of memories, a pageantry of tender moments accumulated over a lifetime, and the unique nexus of human acquaintances that made up the tapestry of a single human life. Each loss is registered in three books with different impacts: the loss to our family, to our generation, to the great flow of life that is the human species. Cosmic death, however, is far worse than any of these.

Who were your father and mother? Your grandfathers and grandmothers? They lived long lives as rich in detail as your own. They wrote letters and took pictures of things that struck them as important at the time. They probably had impassioned feelings about politics, religion, and the general state of the world. They worked, went on vacations, marveled at the sunset and the blue skies, gazing at the very same stars that you now see. Some even wondered about the meaning of the Void. Yet even as little as one life span ago, it is impossible to recover many details within a family, of who these people really were. A few stray photos. A yellowed letter. A handful of interesting anecdotes passed on by word of mouth. We all come from rich family histories that extend many generations back through the corridor of time. We feel a sense of continuity with our own past, but it is a surprisingly fragmentary one at best. Yet when it comes down to the details, we hardly know our own parents or siblings in as much depth as we imagine.

In the years before my brother Leonard died, we had many long conversations discussing spiritualism and the physical world. In my youth, I engaged these conversations with much more passion than I could muster in middle age. It's hard to say what had changed in me over the years. Perhaps it was because our conversations eventually traveled along familiar ruts. Perhaps it was simply the realization that as adults, we could never reach agreement on basic principles. Leonard was deeply convinced that a spiritual universe coexisted with our own. I was not. Death, he said, was the physical transition to full tenancy in this incorporeal universe but our Karma often forces us to return to the physical plane to work out a number of cosmic lessons. "Prove it," I would think to myself, being too polite and sensitive to confront him about his passionate beliefs. He would say there were invisible forces at work that astrologers could read out using planetary positions, and I would counter, "Astrology doesn't work and is wishful thinking." During his youth, Leonard had used LSD and had several near-death experiences. He made annual pilgrimages to an ashram in Pennsylvania to work with his guru to resolve matters of mind and body. Understandably enough, I was never able to offer him any advice or insight since my science had rooted me firmly in the physical world, with all of its distracting rules and regulations. He was the active, passionate traveler on some road to spiritual completion. I was merely a spectator watching him work out his own destiny. Leonard would occasionally lecture me in his good-natured way about the narrow-mindedness of my scientific ways. The words, no matter how tenderly delivered, always stung. He had emotional conviction. I had numbers. I would remind him that science had discovered many beautiful things about the world, exotic patterns buried in the Void that could never have been found in narrow-minded pursuit of the tangible world.

But we loved each other despite our different points of view. I worked hard to understand his beliefs, and he was clever in the ways of physics. In the end, Leonard died alone in the physical world in a hotel room. He passed in his sleep as his liver failed him, but I hope he experienced some of the wonderful transition that he expected as he walked into the light. He left the world with few possessions and no progeny. As he said to me on many occasions, he was content that his physical footprint in time would be overshadowed by his much greater spiritual footprint. In the end, he had hope, whereas I seem to have been left with only the coldness and aloofness of an indifferent universe. It is a universe capable of enormous beauty, filled with rainbows and mystery, yet somehow lacking any evidence that there is anything more to it than meets the eye.

I have often wondered what it might be like to live in both of these worlds with equal conviction. Most humans do so with little effort, it seems. They obey the speed limits and understand nature's limitations by not walking out tenth-floor windows, but they are taught as impressionable children that there really is a spirit, a soul, and so on. Then they learn about science, while at the same time feeling a sense of disdain for the physical world. It is, after all, just a disposable way station to the afterlife. They laugh at the Egyptians, who preserved themselves as mummies: How stupid to expect that their physical bodies will survive eternity. Then they insist on not being cremated because they want their spiritual body to remain intact. While learning about the soul, they also choose whether to believe in other invisible things—hobgoblins, fairies, demons, angels, mystical forces, and the like. How do they decide what to accept and what to reject? Some people accept it all. Others draw a line by only accepting angels and souls, perhaps a few demons just to round out the world of good and evil. It all seems so horribly arbitrary and self-serving when you elect to lose the bonds of logic and let emotion and belief control your thoughts.

On March 22, 1895, in Clonmel, Ireland, Bridget Cleary's body was found in a shallow grave. The coroner said that a sack had been placed over her head while she was still alive. In full consciousness, her abdomen and back had been burned so that her bowels were at last exposed in an incomprehensibly grisly torture. Her tormentors were not demonic criminals. They were her husband and ten of her closest relatives and neighbors, who were convinced she had been taken away by the fairy folk and replaced with a devilish changeling. They simply wanted her to admit it. On a larger scale, over 1 million women were burned at the stake as witches between 1500 and 1700 A.D. in Germany, Holland, and the United States. These people were not killed because of some academic or philosophical misunderstanding. They were the down payment on the price we pay for believing in supernatural agents, unregulated by the simple application of reason and logic. We are powerless to change our ways because in the final analysis, we are compelled to believe these things because of a single overriding fear: the fear of death.

Those of us who have survived to the present moment have a disturbing sense of our own mortality. As we busy our lives with our children and work, we also whistle our way past the cemetery. Look closely into the eyes of your children. In the depths of the dark pupils that calmly regard you, there is the mind of a traveler through time who will outlive you. The landscape of their thoughts at this moment are as unknowable to you as the color of a sunset on a distant planet or the shape of space at the cosmic scale. Yet as you nurture these new beings to maturity and watch them enter adulthood, you feel a sense of satisfaction that some part of you will survive for at least a few decades beyond your own passing. It isn't immortality that we all yearn for. It is just a simple, wistful hope that we will not be forgotten in the vibrancy of the generations to come. You have lived a rich life, filled with wonderful moments and impassioned thoughts. You do not want this life to be forgotten by your grandchildren, though you know that most of it will. The death of the individual takes many forms, the least of which is the ending of the physical form. In ancient Egypt, your final death came when people ceased to utter your name. Pharaohs erased the names of their predecessors from monuments and temples as a final act in obliterating their spirits, denying them immortality. If the death of the individual is disquieting, the death of a community or generation has its own unique twinges of sorrow and finality.

In time, perhaps by the year 2040, the baby boomer generation will vanish just as other generations have before it. The last member of the Civil War generation died in the 1950s; the last of the World War I generation, in the 1990s. The last member of the World War II generation will precede the baby boomers by a few decades. Each generation had its turbulent youth, its young adulthood, its maturity, its old age. As we look at faded family photographs, our father looks back at us as a proud young man standing beside his first car. We see great-grandparents smiling at us from their youth a century ago, posing in their wedding clothes. Each generation goes through the same experiences; only the external circumstances seem to change. Life takes on the aura of one vast party. We enter the party when we are young, dance for a while to the music. We sit down to a good meal and spirited conversation, then it's time to go.

Taken as a whole, the passing of a single individual leaves a complex mark on his or her generation in proportion to the opportunities the individual was able to develop. Most of us, however, will leave no lasting historical record. In London, 7,000 skeletons dating from 1250 A.D. were unearthed at a construction site: a mass grave of humanity. One person's face was reconstituted with clay and hair. In the skilled hands of a forensic archaeologist, the haunting likeness of a young man took shape. From his bones, a synopsis of his entire medical life could be read out as if from a book. We don't know his name, but he was utterly typical—you have seen his likeness before. If you are a builder, your legacy of a bridge, a skyscraper, or even a suburban home may survive several generations. As a politician, an artist, or a scientist, your contribution may survive for centuries. Like the pharaohs inscribing their names on sightless statues, those who write such common graffiti as "Kilroy was here" are calling out down the long corridor of time for someone to be aware of their passing. Since the earliest inscription on an ancient Roman wall, graffiti have signaled the fleeting opportunity to leave a personal mark on history. On a large boulder in Yosemite, I inscribed my own mark that will outlast me. In a strange way, I find it a comfort to know there is such a personal remembrance of me out there. But even inscriptions written in stone are destined to fade with time.

> Only at Thebes can you walk along a track that the ancient people made and still see their scribblings on the cliff, now often with an archaeologist's number penciled alongside; suddenly you are strolling alongside a scribe and his sons, out for a walk in the timeless hills. (John Roemer, *Ancient Lives*)

As we walk through a cemetery kicking the fall leaves swirling about our feet, we come upon markers stripped of any details that tell us who was laid to rest there. Other stones bearing names and dates beckon, silently pleading for us to remember who they were. It has been a century since anyone who knew them visited these people's graves and recalled a humorous anecdote about them. They long for the simple kindness of being remembered. Cemeteries are filled with a dozen generations of humans who have passed or are now passing anonymously into history. This is how it has always been. We do not have to unearth a skeleton from 1250 A.D. or 30,000 B.C. to find an individual whose life has been totally lost to us and to the flow of human history.

In the far future, previous generations will blur into a collection of odd facts that make up the entire pageant of our history as a species. Future generations will look back at our time just as we now look back upon times preceding our own. We uncover the bones of remote ancestors who experienced the hardships of ice ages or the descent from the trees to the savanna. We regain momentary contact with a few stray individuals who made a particular stone ax, or a bone flute, or a cave painting. We see a child in a 70,000-year-old grave, buried with flowers. In the bones of another, the signs of what must have been a painful accident or disease are an open book for us to read. From the distant past of a million years ago, we examine scattered stone implements made by human hands not unlike ours but operated by intellects housed in a brain no longer like our own. In the far future, we see uncertainty and cannot fathom which of our modern trappings may survive to signify the passing of another thousand years of intellectual progress. But just as "Kilroy was here" provides the individual with some measure of immortality, our technological spacecraft castoffs now hurtling past the orbit of Pluto give our entire civilization some solace in the knowledge that these handsome gold and aluminum handiworks will outlast even our own species.

The death of the individual, or the human generation that accompanies it through history, is a tragedy scripted by our DNA. It also makes the art of living that much more precious and urgent. But in the depths of time, long after humans have left the scene, there will come an inescapable Judgment Day. Our Sun must play out its old age according to a script preordained by gravity and circumscribed by its finite supply of nuclear fuel. It must steadily increase its output of energy as its heavy core of helium ash gradually contracts, ratcheting up the temperatures of the nuclear fires. We worry about a rise of a few degrees due to global warming today and the disastrous impact it will have on the quality of life. This is a minor blip in the relentless heating of the Earth to come. In the millions of years that stretch before us, the Sun will steadily heat up the Earth, along with its fragile biosphere. After another 100 million years, there will no longer be ice or snow on Earth even during the darkest winters. After 500 million years, familiar continents will have regrouped. The oceans will reach a new equilibrium with the atmosphere. There will be more water vapor carried by the air year-round as a new level of greenhouse heating comes into existence. Earth will reach a balmy temperature of 90° Fahrenheit during the winter nighttime in the polar regions. Life as we know it will have changed. Many species will have vanished. New heat-resistant species will take their place. Tropical fish will be only fossils, replaced by hardier forms of ocean life we cannot imagine. Warm-blooded mammals will become extinct because it will never be cold. Surplus heat will be a liability for survival. Virtually all forms of life will be reptilian and nocturnal to avoid the desertlike conditions caused by the daytime Sun.

By 700 million years from now, eons before the Sun becomes a red giant, even the biosphere will become extinct. As the Sun continues to grow in power, so much of the water in the oceans will have evaporated into the skies that the Earth will be enshrouded by a thick atmosphere rivaling that of Venus. The surface will sizzle at a temperature of nearly 200° Fahrenheit. All forms of complex surface life will have become extinct. Only heat-loving bacterial forms deep underground will survive, as they have already done for a billion years or more. In time, even they will vanish into lifeless oblivion. At the present moment, then, we are about halfway through the great experiment of multicellular organic evolution on this planet. It has taken nearly 600 million years to ascend from simple bacterial forms to sentient beings. When seen against the broader history of what may lie in the future, the next 700 million years doesn't seem comfortably long enough.

At that future time, alien observers on a distant world, or perhaps future generations of our own species, may turn their instruments to our solar system. They will see one hot world, Venus, with a dense carbon-dioxide atmosphere. They will also see a twin world, Earth, with an equally dense atmosphere of water vapor. They will make the same calculations that we now make as we gaze upon distant worlds in other planetary systems. They will enter a designation for Earth in a table, noting that it is a barren planet orbiting a post-Main Sequence G-type star, now devoid of organic life but which may have had more favorable conditions a hundred million years earlier. They will not know or ever have access to the great fossil record that, like "Kilroy was here," proclaims the history of life on this world. Leaving this solar system and transplanting our biosphere in some future Noah's ark would be our last best hope that at least terrestrial life might survive us as a species. But that may not be our destiny. Even if we do manage to leave our terrestrial home, there is yet another Judgment Day to confront.

As we enter the twenty-first century, astronomers are closer than ever to knowing empirically what the destiny of our universe will be. Encompassed in the shape of the universe and its unseen darkness is a glimpse of its far future. It is an increasingly bleak and fearsome destiny that is already written in the numbers we have found. Not only did a bat and a ball make contact 15 billion years ago, but there was also a wind blowing through the ballpark. That wind will spoil the game. Eternity will not be a friend to matter. If the breakneck cosmic acceleration prevails, in 10 billion years the intergalactic Void will be swept clean of every galaxy we see today. All the stars in our heavens will burn out in a trillion years, replaced by dark embers of collapsed matter. Any free matter on planetary surfaces or in the dark interstellar spaces will also decay by a trillion trillion trillion years hence, ending any future existence for life based on protons and chemistry. Even these dark stellar corpses will eventually evaporate into a brief hailstorm of elementary particles. In the end, in a state of existence mathematicians call "asymptopia," space will have lost all of its interesting luster: its free particles, its black holes, its sentience. It will cease to be experienced. The ultimate death of a universe is not in the destiny of fields and matter, but in oblivion. Even the atoms themselves will no longer survive as our tombstones to carry a record that at one time this universe harbored life. In the end, only the implacable gravitational field of the cosmos, or whatever other fields now lurk within the Void, will continue. Our epoch of life will recede into the unimaginably ancient past as a brief interlude in time—a brilliant epoch of light lasting a mere trillion years, fading into the eternal night. Life dances in this trillion-year flash of light, and is gone. In the meantime, we have our own mortal hopes and dreams nurtured on remote mountaintops or in a brief view through a telescope. We fear death. We fear the darkness it brings. More than that, we desperately hope the grim destiny for a living cosmos that we have discovered in our science will someday prove in error.

Richard Simpson passed away in 1996. He had started out with as much love for space as I had. Together we built telescopes and joined other friends at "star parties," but as we both entered college, it became painfully clear that our destinies had diverged. Math and physics were a struggle for Rich that soon forced him to look elsewhere for career options. He never put aside his passion for space. As we grew older, he seemed stuck in a time warp reliving earlier, more optimistic moments in his life, when the mysteries of the stars were still within reach, before the bitterness of his life had set in. In his mind he may wistfully have imagined a life as an astronomer, exploring where his telescopes could not reach. Whether his view of an afterlife was as well crafted as my brother's I will never know. In the end, whatever satisfaction or uncertainties he found in the darkness he saw between the stars could not support a failing heart that had broken long before.

8

BETWEEN SHADOW AND LIGHT
Quantum Gravity and the Nature of Space

A connecting principle,
Linked to the invisible.
—Sting, "Synchronicity I"

My wife, Susan, and I love to drive up the eastern seaboard to New England in the fall to take in a few days of colorful Vermont foliage and spend time with our family. On one such day in 1985, we loaded up the car at 6:00 A.M. and set off on our usual ride up Interstate 95. As we approached the Baltimore area, the normally bland sky was filled with an amazing spectacle in progress. From every direction, millions of black birds were migrating south. The breathtaking migration flow was creating a pattern of dozens of great tubes of living, flapping, creatures that snaked across the sky 500 feet above our heads. When we got out of the car to get a better look, we could hear only the collective rustling of their wings. Not a single peep or fatigued chirp. Each tube was fifty feet across, with razor-sharp edges. As a bird neared the edge, it would quickly turn back to the interior as though some invisible force were repelling it. The migratory flow moved against a gray autumn sky. The contrast of the geometric pattern against the featureless sky only added to the drama of the moment. From their individual perspectives, the birds couldn't possibly be aware of the geometric shapes they were painting all across the sky. Yet from where we were standing, it was clear that something else was going on that made these living patterns mysteriously emerge through the collective actions of unthinking birds. But what? What could possibly be the instruction or guiding principle that tells a single bird to weave a hundred-mile-long tube in the sky? Do different species weave different patterns, I wondered?

We now understand that there are fields in Nature that, like birds swarming the skies, also paint the Void with shifting patterns of energy, patterns that sing out in the mathematics and data with the same inevitability as migrating birds. As far as we can tell, the fields of matter and force that ply the physical world are like stage actors following a well-rehearsed script, or perhaps like the frostings painted on Nature's invisible cake, each its own color, following the contours of the cake. No matter what kind of particle or field you are, you can only respond to your surroundings in the way laid out in your script, and Nature allows you to play only a small number of distinct roles. Is it possible that there is really only one actor playing many parts on the stage, a single kind of frosting flavored or colored a specific way? Physicists are certain this is the case. Who can blame them? Just think how successful their search for simplicity has been. First, they showed how magnetism and electricity blend seamlessly into electromagnetics; then they found that electromagnetism blended with the weak force to become one. This spirit of research was not imposed upon the world by physicists. Rather, Nature has rewarded physicists by showing, time after time, that the complex things around us are made from simpler things. Even the patterns glimpsed in the morning sky resolve themselves into the simpler elements of birds hurrying to warmer climates with others of their kind in close companionship.

The forces resolve themselves into specific fields built from their own special kinds of particles. To go beyond this level of understanding and ask *why* they do so, we need to abandon searching for clues within the scripts of the three forces. You will not find a clue within the anatomy of a bird that leads to knowing why their collective numbers create migratory patterns in the sky. When an actor reads multiple parts, a good deal can be learned about the play by studying the lines, but there is also the stage and the scenery that shape the actor's actions. Like the birds tracing their living patterns across the Baltimore sky in response to some unknown urge or the frosting hugging the contours of a cake, the elementary fields in Nature also seem to be invisibly guided by some simple principle or law. If there's no evidence for a cosmic Ether or a prior-geometry that Nature's cast of characters can use as a stage, then there is only one other place we can search to find out why things behave as they do. We must look at the gravitational field itself and the Void that sustains it.

It is one of Nature's ironies that the most common force in the universe is also the most miserly in revealing how it works. Gravity is the most important force any of us has to deal with in life. It is also much weaker than any force physicists encounter as they explore the inner workings of the atomic world. The entire gravity of the planet Earth is not strong enough to pull a single paper clip from a toy magnet. This tremendous difference in strength of forces means that our deep understanding of the other forces has flourished, while equivalent lines of investigation for gravity remain completely unexplorable. Something as simple as deciding whether gravity waves are possible or exactly how the gravity force works at the atomic scale has been a major challenge. From the time Maxwell predicted that electromagnetic waves should exist to the time Hertz detected them was a span of only a few years. From the time Einstein predicted gravity waves to the time their traces first appeared in the data required sixty years. Even so, we still can't create them under laboratory conditions. Maxwell explained electricity and magnetism in terms of a single field, and it took nearly eighty years before physicists could figure out how the field worked at the atomic scale. Even shorter lapses of time connected the discovery of the strong and weak nuclear forces with the mathematical language to describe them. The main reason for this breakneck progress in studying nongravitational physics is that the energies where these forces first showed themselves were comfortably within the exploration range of human technology. But this is not the case with gravity.

No matter what physicists do, gravity remains a faceless, unblemished enigma. Physicists can't see the texture of the cake to figure out how it supports the frosting. They can't create gravity fields artificially so that they can be studied. It's hard to explain how something works when there are no guiding details. Still, taking the lead from studies of the other three forces, many physicists continued to explore the foundations of gravity, hoping to coax some details out of the landscape of a mathematical world rather than out of the data-rich physical world. They were looking for the barest logical clues, the merest hint, that might explain gravity's behavior. Many physicists were convinced that once they fully understood gravity, they would uncover answers to some of the deepest questions anyone had about the physical world itself: Why does time exist? What conditions triggered the big bang? Why is there such a thing as matter? Why is space-time four-dimensional? Einstein's general relativity had so wrapped gravity within a pretzel of related issues that to understand gravity, you would have to come to terms with the real meaning of space, time, matter, and field. It didn't take long before physicists realized that general relativity couldn't answer some of the most pressing questions in physics.

The keys to unlocking gravity's secrets didn't seem to exist within general relativity because it was itself an incomplete theory, with no room for the entire quantum world. Attention therefore turned to the one arena in which physicists had scored their most recent and spectacular successes: the world of quantum fields. There had to be a pattern that went beyond the three forces they had studied so intently, a pattern that was big enough to include gravity and all of its trappings. But what could it be? For decades physicists did the only thing that seemed to make sense to them. They struggled onward as best they could, with only a sketchy map of gravity's landscape as a guide. They had no data to lead the way. Only a handful of tantalizing patterns scattered among the workings of the other forces seemed to hold clues to something deeper. They had to fall back on selecting good and bad theories for gravity based on logical consistency alone. Every year or so, new ideas cropped up in the mathematics like dandelions on a spring lawn, as the search itself proved to be rich in new ideas. The ideas and problems physicists kept running up against were so complex that even beauty and logic sometimes didn't seem enough to keep them on the right path. As physicist Stanley Goldberg reflected in 1984, "The quantization of the Einstein theory . . . raises fundamental questions about the meaning of geometry, of space-time, of manifolds, and of the relationship of gravitation to the rest of physics."

It's very hard to imagine trying to create such a theory. If you were to succeed, it would probably overturn even your most basic intuitive beliefs about the very nature of the physical world, especially space and time. Einstein had shown there were intimate relationships among gravity, space, and time. It seems almost incredible to imagine that at some level, gravity's field (space-time) is as grainy as beach sand, just as the other forces are known to be. Objects dissolve away into disconnected bits of space like the image in Plate 5. It is bad enough that the vacuum had to be populated by virtual particles or filled with the shadowy Higgs field. If the entire physical world that they occupy is based on a foundation for space-time as changeable as the shifting sands on a beach, then our very notion of a fixed time and space vanishes. If you were a jellyfish, you wouldn't find a world that fluid much of a problem. A jellyfish is barely more substantial than the medium it lives in. Because we live on a solid planet with rigid buildings and scenery, humans have grown used to thinking that space is somehow just as rigid. But it isn't. Einstein tried to take the pulse of this research in the 1930s, but the prospects seemed pretty dismal since there wasn't anything to anchor the search to: "It is not unimaginable that human ingenuity will someday find methods which will make it possible to proceed along such a path. At the present time, however, such a program looks like an attempt to breathe in empty space."

Within a few more decades, physicists were trying many different approaches to this "breathing exercise," in their efforts to scale the steep "mountain of mathematical abstraction." This wasn't day hiking on an ordinary rocky promontory of merely difficult mathematics—permits for this climb were reserved for only the most mathematically adept. That left everyone else in physics and astronomy at the base, scanning the peak to catch a fleeting glimpse of the incredible vista above. Occasionally, a few tired theorists would stumble back down, shaking their heads, and rest up for the next traverse. They spoke a bizarre language, and to understand what they were saying, you had to master the mathematical language and accept their visions of space and time.

If it turns out that gravity looks like the other fields and that those fields are textured because they are quantized, then gravity will also have to follow the same pattern. What does it really mean for gravity and space-time to be grainy or quantized the way the other fields are? It means that space and time are not infinitely divisible into smaller things. At some physical scale, space and time cease to exist and blur together into some poorly defined new entity. Since the early 1900s, physicists have felt certain they know where and when this graininess might start to show up. The scale is unimaginably small. By combining the three key natural constants in Nature—Newton's constant of gravity (G), Planck's constant (h), and the speed of light (c)—physicists can create units of a fundamental mass (0.000055 grams), length (0.00000000000000000000000000000002 inches), and time (0.00000000000000000000000000000000000000000014 seconds). Although physicists did not have a verifiable theory that combined general relativity and quantum theory, the combination of these constants that are the cornerstones of quantum mechanics (h), relativity (c), and gravitation (G) seemed to hint at a glimpse of a more complete theory. This isn't much more than a kind of mathematical numerology done on the back of an envelope, but it is almost universally accepted by physicists that something very weird probably happens at these unimaginably small "Planck scales" in the physical world.

The problem is, no one really knows for certain what it might be.

Imagine an atom the size of the entire Milky Way galaxy (100,000 light-years, or about 600,000 trillion miles). In this model, the Planck scale would still be smaller than the size of the period at the end of this sentence. At the present time, physicists can probe the structure of the atomic world at 0.00000000000000000001 inches. For our Milky Way–sized model, this is about the same as seeing things as big as Earth's orbit around the Sun. It's a dreadfully long way from the size of the orbit of Earth to the size of the period at the end of this sentence, where, in our scale model, space might somehow break

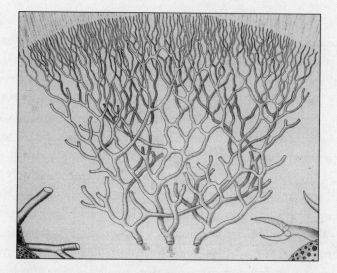

FIGURE 8.1   Empty space dissolves away into an interconnected network of energy embedded in the cosmic gravitational field. The shape of a fossil radiolarian seen under a microscope by the nineteenth-century microscopist Ernst Haeckle mimics the flow of large-scale form into a finer-grained labyrinth of ever smaller structures. (Courtesy of Kurt Stueber, http://www.mpiz-koeln.mpg.de/~stueber/stueber_library.html)

apart or become fragmented. There could be a lot of hidden detail in the staggering gulf of scale in our model between the orbit of Earth and that period—detail like that shown in Figure 8.1, where an ancient radiolarian shell explodes into detail under the microscope. If gravity follows the grainy patterns present in the other fields, it will be hard to prove this if the Planck scale is the magic key.

Aside from being a grainy field, gravity may also share another feature of the other forces: Nature takes a dim view of mathematical exactitude. There are specific limits to measuring things in Nature, especially atomic things, and these limits permit a vibrant, scintillating virtual world to flourish in the Void. In the 1930s, Heisenberg discovered a crucial cornerstone in quantum mechanics: Nature prohibits us from knowing both the exact speed of a particle and its exact position. It also refuses to let us measure a particle's exact energy at a specific instant in time. The same restrictions also carry over to descriptions of fields. You can't tell exactly what a specific field is doing because the details are hidden in the comings and goings of virtual particles that make up the texture of the field. It is this fuzziness that lets Nature fill the Void. This is such an overpoweringly important idea in quantum mechanics that almost all physicists believe it will play a big role in quantum gravity theory. Because gravity fields and space-time are firmly linked together in relativity as one and the same symbol, when this quantum fuzziness is applied to gravity, space-time cannot be a rigid thing. Space-time can't be a static arena on which Nature plays out its dance like the atoms in a crystal in Plate 6. Instead, like interfering ripples on a pond that toss leaves and bugs around, our world is the result of many possible alternatives for space and time that blend together at any given moment, like the elements of a dream emerging from individual neuron firings in the brain. In quantum gravity, the precise knowledge of both the shape of three-dimensional space and the speed at which it changes in time is also forbidden. Space-time only makes sense if you use averages, and at sizes much larger than the Planck scale. This also means our entire four-dimensional space-time, defined by the worldlines of individual particles, has a distinctly fuzzy shape at the Planck scale, kind of like reading the morning newspaper with the wrong pair of glasses.

What does this fuzziness mean to you and me? Practically speaking, it means nothing at all. There isn't a single thing we experience in Nature or can easily measure that will be affected by whether or not gravity is quantized. Particles may jiggle around, but it is at a scale a billion billion times smaller than the nucleus of an atom. But intellectually, and speaking from the heart, the implications for our intuitive notions of time and space are nothing less than catastrophic. If space and time can at some point lose their individuality and certainty, how do we then regard all the other possibilities for space-time that might have been? What becomes of the sanctity of cause and effect, or the sense of self, when the bedrock of reality seems to tremble and shift from state to state in an uncontrolled and chaotic dance? In the quantum world, things remain in a half-real state until they are observed. Does this also mean that space-time doesn't exist until it is observed? And who is doing the observing?

Leaving these philosophical issues aside, there are also some annoying technical details to address. If we want to make gravity fit into the pattern of the other three forces, there has to be a particle, a quantum of energy that transmits gravity—just the way the photon transmits electromagnetism, the gluons transmit the strong nuclear force, and the W/Z particles transmit the weak force. If that's the case, this "graviton" particle must have some very specific properties of its own. First of all, gravitons can't have any mass at all. If they did, light from distant stars seen near the limb of the Sun would not be bent in exactly the way that is observed. Gravitons must also have exactly twice the spin of the photon, the gluons, and the W/Z particles. If they didn't, the mathematics says you will end up with equations that don't look anything like Einstein's equation for the gravitational field. You wouldn't get big bang cosmology or black holes, nor would you have any of the other rock-solid tests of how gravity works at the largest scales in Nature. And finally, there is one more thing to consider before we have collected the main components of this quantum gravity theory. What do we do with the other particles and fields that are actually creating the gravitational field? With your mind's eye, imagine a line drawn on the fabric of space. Where do the points on that line stop being space, to become instead the embedded quark, electron, or gluon? How does a gravitational field "become" an electron or a quark, which then turns around and generates gravity and space-time in the first place? At some point, to understand gravity we have to fully understand how it would be produced among objects that are themselves quantized. There will also have to be a common description of gravity that is shared by the other particles and fields in Nature. At the core of this new description, we have to confront the baffling idea that matter is actually a form of space.

It is not easy to imagine that matter could be just another form of empty space. We are, after all, still struggling with Einstein's nearly one-hundred-year-old idea that mass is just another form of energy. The very idea that matter is a form of space has profound implications that reach the very bedrock of physical reality. Toward the latter years of his life, Einstein was so certain that matter was not really a fundamental ingredient in nature that he wrote: "The material particle has no place as a fundamental concept in field theory. Even Maxwell's electrodynamics are not complete for this reason. Gravity as a field theory must also deny a preferred status to matter."

Einstein wasn't the first person to wonder about this in recent times. In the eighteenth century, George Reimann and Clifford Will also thought matter was somehow just a by-product of the geometry of space. The theoretical program of actually trying to convert matter into space didn't really get started until the German-American physicist Hermann Weyl (1885–1955) published his 1924 paper "What Is Matter?" This theme was adopted by Princeton physicist John Wheeler (1911–) and University of Maryland physicist Charles Misner (1932–) thirty years later. By that time, it was becoming clear what had to be done to make such a program come into its own. To even begin to make such a radical idea work, space-time at the Planck scale couldn't just be smooth, like a flat sheet of paper. Because the quantum field of gravity was fuzzy and grainy, its geometry had to be complicated and frothy. Wheeler thought that within this complex frothiness it should be mathematically possible to create a description for matter. Somehow, in the mathematics of the quantum foam, there should be some relationships rigid enough to serve as the foundation for an electron, a quark, or a photon—an ultimate vindication of Weyl's dreams. But this approach brought a monumental disappointment. By 1972, Wheeler had all but completely given up this particular quest. No matter what exotic mathematics he tried to use, no rock-solid relationship ever emerged that looked like a recognizable particle. Science popularizers, meanwhile, still use this "foamy space-time" image whenever they write about gravity, as though it provides an authentic glimpse into the unknown.

The problem with a foamy space-time was that there didn't seem to be any way to ensure there would still be fixed relationships in the geometry that could be identified as electrons or photons or anything else. No matter how it was modified or how foamy the mathematics became, Einstein's general relativity just didn't provide a natural setting for a very specific and crucial atomic property: spin. As already mentioned, spin helps physicists keep particles and fields separate in the mathematical ledger book. At the same time, it is a property that has virtually nothing to do with our image of a toy top or a gyroscope turning on its axis; it has more in common with the biological concept of sex, which only takes on two states, male and female, although of course sex has no counterpart in the atomic world. Particles that carry a spin of one-half a quantum unit include electrons, neutrinos, muons, and quarks, which are often just called fermions. Particles with a spin of one unit include photons, gluons, and the W and Z particles and are called bosons. As for particles with a spin of zero, the only ones available to us seem to be the Higgs particles, which haven't really been discovered yet. Meanwhile, gravitons are the only particles that physicists know about with a spin of two units. They have never been detected, but the mathematics states very clearly that they are the end of the line in our universe. There are no other fundamental particles with three, four, five, or more units of spin. If we were to find any, they would cause a major upheaval in physics.

The mathematics of general relativity is so complex and subtle that it took over fifty years before physicists like Wheeler had explored enough of it to declare that spin doesn't exist within its logical fabric. Without having spin as a means of sorting out one particle type from another, there could be no mathematical way to distinguish electrons from quarks, or photons from gluons, in the equations. You would have a theory of pure gravity and gravitons, but you could not locate where the familiar particles of our world were supposed to be in it. It would be like making a cake and forgetting to add the frosting. No matter how hard you stare at the cake, all you see is cake. There would have to be something buried inside general relativity's model for space-time that allowed flexibility on the largest scales but was absolutely constant at atomic scales so that the properties of fundamental particles could be preserved, not only on Earth but across the light-years of cosmic space, like the galaxies forming filamentary patterns in Plate 7. In some sense, general relativity was incompatible with quantum mechanics. Although general relativity claimed that matter causes space-time, there wasn't a single property of atomic matter that could be found hidden in the mathematics of the theory. So long as physicists stuck to Einstein's simple geometric model for the gravitational field, there could be no other particle in the universe than the graviton itself, cut from the very fabric of the gravitational field. Because general relativity denied a natural place for matter in its description of space-time, it was a puzzle that lacked a picture. The pieces were all there, but after assembly only a blank, featureless Void stared back. As long as you insisted that fundamental particles were points of energy with no size at all, there wasn't a whole lot you could tack on to them. But what if they weren't points? What if they were something just a little bit more complicated? Could that provide enough leverage to pry open a new line of mathematics that could unify everything?

In 1984, John Schwarz at Caltech and Michael Green at Queen Mary College announced something called "superstring theory." It was actually based on a collection of ideas nearly twenty years old by the time they dusted it off. At one time, physicists tried mathematically to stick a quark on each end of a string of energy to make some of the nuclear particles. Once gluons came into vogue, the string idea vanished faster than a neutron decaying in the night. That, plus the fact that when this earlier theory was worked over, its creators found that these strings preferred to live quietly in a twenty-six-dimensional universe; otherwise they tended to spawn impossible "faster-than-light" particles called tachyons that could turn around and kill their grandparents. Now a new use for strings appeared. General relativity said that space-time was too barren to account for quantum mechanical rules, so there was something missing from a world of point particles and worldlines. String theory provided physicists with a new mental and symbolic image for matter and field that they could now use to bridge the gap. With strings, particles would not be thought of as pointlike knots of energy but as equally mysterious one-dimensional vibrating strings of energy that flowed through space-time in macaroni-like worldtubes. It is very hard to imagine what these strings are supposed to be in terms of any analogy from our familiar world. Like the idea of spin, the term "string" is just a placeholder for some new property of Nature outside of human experience. In string theory, every particle is its own loop of vibrating "somethingness." Whether it represents a graviton, a photon, or an electron depends on the specific way it vibrates. Electrons that look like some kind of diminutive loop of energy have never been seen because stringiness doesn't show itself until you reach the Planck scale. At least that's what a quick back-of-the-envelope estimate seems to suggest.

I imagine strings as being some kind of microscopic, luminous loop that whizzes around, but that's not really what they are supposed to be, either. Only string loops that have the properties of photons are actually luminous. The rest of them represent particles of matter that move and vibrate more sluggishly in the darkness. Every virtual particle we have already talked about that inhabits the Void is actually one of these little loops of energy scurrying about behind the scenes. It's a funny thing to realize that despite the hundreds of "explanations" of what strings might be, some by the creators of the theory themselves, no one has ever explored how to think about them or how they are related to other things in physics. For example, if a string can be a photon, and a beam of photons can be produced by your flashlight, does that mean your flashlight is some kind of rapid-fire machine gun that rips these loopy strings out of the vacuum, ejecting them into the real world? If the wavelength of a photon in the radio region can be over a yard long, does that mean that one of these string-photons is in some way a yard long? If that's the case, how can a string-photon that is only 0.000000000000000000000000001 inches long have information in it about how a radio wave photon is supposed to behave, one that is as long as the line of text on this page?

It helps to think of these strings as being like violin strings. The lowest vibration is actually what the physicists call the zero-mode. Paradoxically, even though it is not vibrating, a particle that looks like a zero-mode in string theory can still carry energy as it moves through space-time. A physical example would be a photon that has no mass of its own but can still zip through space at the speed of light and carry lots of energy. The next-highest string wave vibration is a single wave that just fits along the length of the string, much like what you get when you pluck an unfretted violin string. The vibration carries a staggering amount of rest energy equal to 10 billion billion times the mass of a proton. The next-highest—the second harmonic—carries twice this incredible rest energy, and so on.

All of the familiar particles that physicists have discovered are actually jammed into the lowest string that is not vibrating. String physics says that there is not much difference between photons, gluons, quarks, and leptons. They are all just one kind of string vibrating at its lowest mode in space-time. The differences among them are actually very minor when seen against the full spectrum of string vibrations, which extend like a ladder to infinity. This means that if all the particles are on the lowest string, none of them can have any mass at all. This also means the forces they produce must be nearly identical. In one fell swoop, string theory has unified the strong, the weak, and the electromagnetic forces, joining them with the quarks and leptons as nearly identical but massless particles.

How do string theorists *really* think about space-time after the 1982 revolution? Brian Greene, in his book *The Elegant Universe*, tries to explain it. The bottom line seems to be that if you believe in string theory, there is only one logical possibility for thinking about space-time, at least for the moment. If you try to imagine what a quantum theory of gravity would actually mean physically, the idea that space-time is a smooth field completely dissolves away into a pattern of disconnected strings: "But in the raw state before the strings that make up the cosmic fabric engage in the orderly, coherent vibrational dance we are discussing, there is no realization of space or time. Even our language is too coarse to handle these ideas, for, in fact, there is even no notion of before."

The strings themselves are truly fundamental. Even defining what strings physically represent gets us smack into some thorny semantic problems. If you try to imagine what they are, you will immediately see them as constructed of something even more basic than the string itself, which is not allowed by the logic of the mathematics. And you have sworn that this logic will be your faithful guide by accepting that strings mathematically exist in the first place. It is like trying to define what you mean by a mathematical point or a piece of sausage without providing any context like a line or the filling itself. If you think of strings as pieces of interconnected macaroni, you get lost in empty speculation about the composition of the macaroni. String theory says you *must not* ask that question because strings are so basic they are not made from anything at all, yet they have a specific logical form and character that can be deftly written down in an equation. Humans don't like, nor do they accept, the idea that something can be final, that there isn't always one more "Well, what is *that* made of?" question that can be posed without any embarrassment. Nature has provided us with the ability to define things in specific contexts. If you take away all contexts of space, time, matter, and energy, we are helpless. Nature has also given us examples of things that behave the same way, as with spin, for example. We can use the name, but we cannot ask what it is in terms of other things we know about in our world at large. You can ask what some things are, but sometimes Nature will not or cannot offer a satisfactory answer. The answers are literally beyond logic.

In exchange for forcing us not to ask the ultimate question—what they "really" are—string theory does a lot more than just give us a new way to think about particles. It forces us to rethink what space-time looks like and to contemplate exactly where the properties of space and time enter the picture in the first place. You see, it may not be correct to think about the strings as moving through what we normally think of as space-time. This is because space-time is actually the thing that strings produce as they vibrate. They are sometimes the kernels for the gravitons in the gravitational field. At other times, they are the photons or quarks that appear elsewhere. These strings also create the virtual particles that come and go in the Void. Just to tantalize us, here's what Michael Green offers as an explanation of what this might mean, physically:

In a theory of gravity, you can't really separate the structure of space and time from the particles which are associated with the force of gravity . . . The notion of a string is inseparable from the space and time in which it is moving, and therefore if one has radically modified one's notion of the particle responsible for gravity, so that it is now string-like, one is also forced to abandon at some level the conventional notions of the structure of space and time . . . at these incredibly short scales associated with the Planck distance . . . A lot of the present research is focused on trying to understand precisely how [this] works.

But how do strings "do" anything? How do they move? How do they animate matter? I have always had trouble with questions of this

FIGURE 8.2   Mathematical fractal patterns that have no physical existence also mimic the flow of information and structure from large scale to small scale, eventually filling every nook and cranny of space. Unlike real space, however, fractals continue indefinitely into the submicroscopic realm of the infinitely small. (Fractal art courtesy of Linda Bucklin)

sort, because I keep stumbling over Zeno's paradox, which says that after you reduce the universe to a bunch of points, nothing should ever move. If you accept the idea that by their logical construction, strings are derived from space-time, you are close to understanding how they work as elements of the real world. They don't exist as kernels of pure space. They are kernels of both space and time. They don't move at all. They simply *are*. In some ways, they are like the beautiful fractal landscapes in Figure 8.2 that exist in a static world of their own. Like the foamy shapes in Plate 8, they are motionless but have shapes that imply motion.

The main reason these strings work to unify gravity and everything else is that there is another idea absolutely at the core. That idea was discovered in the mid-1970s, when physicists were busy creating models that they hoped would unify the strong, the weak, and the electromagnetic forces all at once. What they uncovered in their equations was a spectacular new way not just to unify the forces in Nature, but to unify them with the very particles they act upon. It's called supersymmetry. Through the microscope of mathematics, physicists discovered that it was possible to treat electrons, quarks, neutrinos, gluons, photons, and the W and Z particles as simply different versions of each other. What does this all mean? Well, let's consider a little analogy between supersymmetry and a pocket full of loose change.

Imagine that you have just completed a tour of several foreign countries. Your pockets are heavy with a variety of coinages from Germany, Sweden, France, and the United States. The total value of all the currencies is a single number representing a fixed amount of gold that these different currencies are worth. In supersymmetry theory, this gold represents the magnitude of the superfield. Under a supersymmetry change, any field can be rotated so that its magnitude remains the same, but the various components of the field have to change in a well-defined way. In our coinage analogy, by going to a bank, we can perform a currency conversion on each of the coinages to change them into one of the other currencies. We can do this without changing the total value of our money in terms of its value in gold. Likewise, the electron, the quark, and all the rest of Nature's fields are the coinages of Nature and can be changed one into the other by supersymmetry.

In addition to the miracle of supersymmetry, there is another important reason that superstrings help to unify gravity with the other forces. Superstrings don't live in Einstein's four-dimensional space-time. They live in a much larger arena with ten dimensions. Within these extra dimensions, Nature does its bookkeeping to decide whether one string is a quark, a photon, or even a graviton. You couldn't see a complete string even if you tried, because a big part of what they are is rolled up into separate, invisible "mini-universes" that have a shape. Because these shapes are endlessly duplicated everywhere in four-dimensional space-time, a gluon or a quark on Earth works exactly the same way as one on Alpha Centauri.

As exciting and mysterious as such a hyperdimensional world might be, there may be other ways of looking at the same mathematics that make it a lot less outlandish and, to some physicists, a lot more palatable. Five years after the revolution, some physicists didn't see that this was the only way to interpret the difficult equations of string theory. Steven Weinberg, for example, has another way of looking at this new landscape: "[O]ne thinks about the theory formulated in four dimensions but with some extra variables which can, in some cases, be interpreted as coordinates of extra dimensions, but needn't be. In fact, in some cases, cannot be."

These extra dimensions may not have direct physical meaning at all, at least not like space and time do. There are three flavors of space (up, down, and sideways) and one flavor of time. Strings require six more flavors of "something else." The color of a quark is in some ways a "dimension" of the particle, but certainly not one you can measure in inches or miles. These extra dimensions may only serve the bookkeeping needs of the theory and may not have anything to do with space as we intuitively understand it. When you divide one number by another in long division, all of the in-between numbers you generate are merely the means to an end. They are thrown away with the scratch paper once you have arrived at the answer to the problem. You can think of them as the hidden dimensions of the math problem. Computers don't even generate these numbers. They electronically jump to the answer in virtually a single step. Could it be that these other dimensions that string theorists need are just tools for human calculation? Could it be they don't actually have a physical existence? In the world of pure mathematics, it is sometimes hard to tell which things have a physical interpretation and which things are merely scratch paper. This ambiguity leaves another avenue of string theory open for exploration.

Strings seem to "move" against a background space that helps physicists keep track of them in the mathematics, but this background space is just a scaffolding that is thrown away later on. I am reminded of a type of wedding cake popular in Sweden called a *spetikaka*. It is created as a lacework of meringue on a conical form. Once the meringue has been baked, the paper cone is removed and you are left with a beautiful filigree cake with an impossibly delicate lacework suspended in midair. This cake could never be made without the disposable form. Strings should be thought of like that. By the mid-1980s, some physicists such as James Gates at the University of Maryland had actually found a way to rebuild string theory so that it was only a four-dimensional theory right from the start. The rest of the dimensions would not be like familiar space at all.

> I am a believer in Sagan's Lemma, "To be accepted, extraordinary theories require extraordinary evidence." Although most of the string community seems taken with the idea of extra dimensions, I remain unconvinced. To some degree I am a heretic in the church of superstring theorists. It may turn out that the now popular higher dimensional interpretation of superstrings will be like the epicycle view of some astronomical observations.

After fourteen years of effort, these versions of string theory have not attracted as much attention as the now "standard" ten-dimensional theory. In fact, string theory seems to have survived this technical challenge and has even passed through a second transformation in 1994 after finding itself in a kind of mathematical quagmire for several years. The revival of interest in the older interpretation of string theory was triggered by a spectacular discovery within a community of physicists who by this time had already created no fewer than five different string theories to choose from. Which one was correct? The answer, to everyone's surprise, was that the five distinct versions uncovered in the 1980s were actually part of an even bigger idea: M-theory. The resolution resembled what would happen if a group of physicists wandered around in a chicken-packing plant and found a leg here, a thigh and a breast over there. Each would seem like a unique object. But M-theory put the pieces together into a whole chicken. If we listen carefully, this particular chicken may tell us the secrets of the universe.

To make sense of strings in *this* theory, physicists had to invent an even more exotic arena in which the strings could operate. Again, to get there, all they did was follow another uninvestigated pattern in the mathematics. The five types of superstring theory that were known through extensive mathematical exploration in the 1980s and 1990s each bore a similarity to the others, but they had important differences. There were also some very peculiar relationships among the many different string theories. One kind of string in one version of string theory could be made to look like another kind of string in one of the other four versions. The clue to seeing what was lurking behind them all required a model in which strings were not fundamental in their own right. Strings were a product of a new, still more enigmatic object called a membrane.

We all know about membranes—most of us spent hours playing with soap bubbles as children. A garden hose can be made from a two-dimensional sheet of plastic (a membrane) that is curled up along a third dimension to create a tube. When you look at it from far away, the garden hose looks like a string. It was something like this that Princeton physicist Edward Witten was thinking about when he suddenly understood that all five superstring theories were actually part of something even bigger. He had found an intact chicken in the poultry factory. There really was a pattern lurking behind the five different superstring theories. It was such an incredible discovery that many researchers thought that "M" in M-theory also stood for "magic" or "mystery." In M-theory, eleven-dimensional space-time can contain islands of three-dimensional membranes, which represent our universe. Other "three-branes" can also inhabit the Void and represent other universes that do not share any of our points in space, not even if these universes were themselves infinite. After all, infinity is so big that you can pack an infinite number of infinite things within it without any of them touching. The strings that are supposed to represent quarks, leptons, gluons, and the like, begin and end on our own "brane." Gravity is different than everything else because it is free to leave our brane and become a part of the larger eleven-dimensional space-time just around the corner. In fact, gravity *is* this underlying space-time. If all of the gravitational force were confined to our braneworld, it would be a force as strong as the others we know so well at the atomic scale. According to Lisa Randall and Raman Sundrum at Princeton University, although the size of the three-branes is infinite just like our universe seems to be, there could also be other large dimensions tacked onto them. The extra dimensions in each of these brane universes wouldn't be infinite. They could actually be as big as our solar system or as small as the period at the end of this sentence. Amazingly enough, they have no effect on the three big dimensions we already know about. Like the ends of a piece of string glued to a piece of paper, the familiar particles and the three nongravitational forces are restricted by string mathematics to exist only within a particular brane. This means there can be no communication or particle exchange between neighboring brane universes. However, gravity is not restricted to a single membrane. It allows different brane universes to interact only by their gravity even though these other, infinite brane universes may only be millimeters away from us along one of the other dimensions. Light from the stars in one universe cannot travel across the higher-dimensional rift to be seen in the other universe. This universe would be totally invisible. Only the gravitational forces from stars in these other branes would affect us. They might even account for some of the dark matter or vacuum energy that astronomers can see in their inventory of the universe.

When you look out at the sky at night, you are seeing a very short distance into these other dimensions, but there is nothing there. The light we need to carry information to form images boomerangs back to us in our own space. It never connects up with the other brane universes in the Void to form any images of what is there, no matter how close these other braneworlds are to our own. So we just see a black emptiness. If we could see by using gravity, however, the darkness might be filled with looming ghostly figures of stars and galaxies in other brane universes. The way that gravity is changed by the number of membrane dimensions is also not random. It is tightly controlled by the mathematics. If there is one extra dimension in a membrane beyond the three we know about, gravity will act very differently at scales of a few hundred million miles. If you add two additional spatial dimensions so that our universe is now a "five-brane," then significant differences for the way gravity works will appear at a scale of only a millimeter or so. In a "six-brane" world, the scale of the extra dimensions will shrink to subnuclear size. Once again, the familiar particles and nongravity forces are trapped into the space-time of the six-brane, like villains trapped in the Phantom Zone in the *Superman* comic-book series.

How do we go about testing such difficult ideas? The most important test actually comes directly from astronomy. A star that explodes as a supernova sheds an enormous amount of energy. If there were other dimensions that gravity could flow into, leaving our four-dimensional space-time behind, the energy released by a supernova would leak out into these other hidden dimensions by gravity waves. The energy balance for the supernova in our space-time would not work out. There would be missing energy to account for in the aftermath of the explosion. When Supernova 1987A exploded in March 1987, physicists detected neutrinos from the detonation and were unable to find any "missing energy" that could have escaped into other dimensions. Everything they saw could be accounted for very neatly by the energy of the event estimated from our own space-time. If Nature was indeed taking advantage of M-theory, it was being subtle in the traces it would permit astronomers or physicists to see of it.

Careful studies of gravity under laboratory conditions will soon be able to test for these extra-dimensional effects without having to wait for another supernova. Gravity can only be tested very accurately on solar-system scales. In the laboratory or atomic domain, it represents a state-of-the-art experimental problem. This testing is already good enough to prove we don't live in a four-brane universe. But tests of gravity at the millimeter scale that would rule out a five-brane world are fiendishly difficult. At Stanford University, Aharon Kapitulnik and his colleagues are trying to set up an ultra-precise measurement of gravitational forces at millimeter scales to search for any signs of extra dimensions. They originally planned to measure other atomic forces as well, but they have abandoned these studies in favor of searching for exotic gravitational effects instead. Who can blame them? Now that superstring theory has opened up a portal into an indescribable world of multiple dimensions and membrane universes, experimenters have been turning away from the more mundane tasks of precision measurement to search for traces of other dimensions in the gravitational field. They expect to fail. But if they succeed, we are in for some dramatic intellectual surprises on a par with the discovery of life elsewhere in the universe.

Another exotic benefit of these additional brane universes was uncovered by Keith Dienes, Emilian Dudas, and Tony Gherghetta at CERN in 1999. If one of the extra dimensions were as large as

0.0000000000000000000001 inches, which happens in a six-brane universe, the three nongravitational forces would merge together, not at the fantastic energies of the Planck scale but near the energy where the electromagnetic and weak forces merge. Physicists should be able to see this happen once the Large Electron-Position Accelerator (LEP II) machine starts working near its maximum energy in 2005. If they see nothing, then M-theory can still push the limit even closer to the Planck scale by proposing we live in a seven-, eight-, nine-, or ten-brane universe. Then we literally run out of brane dimensions in eleven-dimensional M-theory. At some point, M-theory and branes become untestable because our technological resources will no longer let us carry out the needed tests. We have to hope that by then, these ideas might have been disqualified on the basis of some as yet unknown logical flaw. They could also be replaced by yet another idea that would replace strings with an even more potent way to comprehend forces and matter.

A number of physicists have looked at the difficult task of unifying gravity with the other forces and have concluded that perhaps they can never be joined. Richard Feynman, one of the developers of quantum electrodynamics, took an increasingly dim view of gravity unification, questioning whether the incredible difficulty of taking that last step was a clue that gravity is simply different from everything else. Perhaps it really isn't a force at all and can never be described by quantum theory. If this is the case, then gravity must be treated absolutely separately from other things in the world, and string theory's elegant mathematics is an approach to the world based on the wrong starting premise. Superstring theory, and the M-theory it has morphed into, continues to draw new generations of young physicists to its shores. Many of these physicists have read science fiction and have watched *Star Trek*. They have a natural can-do vision of space-time as a malleable fabric woven in multiple dimensions. Yet despite the seemingly endless theoretical excitement that has come out of the font of this theory, its detractors, such as Boston University's Sheldon Glashow, have challenged the entire string community to confess that their theory is in fact hopelessly untestable. Worse than that, Glashow says it is not even a true scientific theory at all because none of its most important elements can be observed or verified. When you try to falsify the theory, its practitioners simply move the goalposts into some invisible dimension. Glashow got so tired of hearing string theory "babble" at Harvard University that he left Harvard and moved to Boston University in 1998, where he set up a more experimentally minded physics group.

Staunch supporters of string theory gladly admit that it probably works best in a world defined by the unreachable Planck scale or at energies trillions of times higher than any likely to be reached by accelerators in the near future. But they resent the idea that it is untestable and unscientific. In QED, physicists had to accustom themselves to virtual particles that made many kinds of calculations and predictions possible in the first place. String theory has a lot in common with such unobservable phenomena, but most of its practitioners are convinced that string theory is far more than just idle mathematical speculation. Physicist Gordon Kane at the University of Michigan says that string theory can actually make many predictions that are now, or will soon be, testable without having to go to the extreme energies where they naturally operate. This will happen just as soon as physicists master its complex mathematics and start to make more predictions testable at the energies we can now reach within the first rung of the string energy ladder.

String theory's greatest prediction so far is actually a mathematical holdover from the 1970s that was absorbed into the fabric of the new theory in the 1980s. Supersymmetry, which ties quarks and leptons to the photons, gluons, and W/Z particles, is a new pattern ready for discovery. This new principle should begin to be seen at the energies near the limit of the new LEP II machine. Because of the way this new symmetry works, there must absolutely exist a whole new spectrum of "super partners" to every known particle. These particles will probably have masses many times higher than the top quark, which is itself nearly as heavy as a gold atom. One particle called the gravitino may have very little mass at all and might even supply some, or all, of the dark matter observed by astronomers. In 1998, researchers at CERN, using the same machine as that used in 2000 to search for traces of the Higgs particle, tuned the twenty-seven-kilometer LEP I accelerator to collide electrons and positrons at energies of 189 billion volts. For years, they had used this accelerator at lower energies to create the elusive tau lepton—the second cousin to the electron. The numbers that were produced fell into line with their predictions, but now this began to change at the higher energies. Instead of detecting about 170 taus, the machine spewed out 228. Because supersymmetry predicted that there should be new partner particles to all the known ones, the higher numbers of tau leptons made much more sense if supersymmetry was included in the calculation. This tantalizing evidence that supersymmetry might really exist was also swallowed up along with the Higgs particles, as the LEP I accelerator was turned off in November 2000 to begin construction of the new LEP II accelerator.

String theory also allows new, but rare, reactions to take place among familiar particles, reactions that are not currently allowed by the other three forces. String theory predicts new kinds of states for matter, which could show up in such unlikely places as the cosmic fireball radiation left over from the big bang. String theory will also let us predict how much dark matter there should be compared to luminous matter in the universe, which is currently split twenty to one. When physicists began to explore the details of superstrings, they were amazed at how many features of the physical world it could explain, even though these features had not been included in the mathematics when they first set it up. According to James Gates:

> For me the essential esthetic of superstring theory consists of a few beautiful simplicities. Superstring theory says that matter and energy are both aspects of a unity, not different things. This is a theme that has been playing itself out in physics for at least half a millennium. Sorting all of this out I fear will take a very long time, but I believe this will occur by paying more attention to a dictum by Copernicus, "Let only geometers enter here." That was the route of Einstein too as he gave us a view of the 4-D geometry of our universe.

So in the simplest terms, at the start of the twenty-first century, Einstein's geometric theory of space-time and the fledgling mathematical theory of quantum gravity both lead us to a dramatically new vision for space, space-time, and the Void. It may be an eleven-dimensional object, with four of its largest dimensions available for direct inspection. It could be a network of stringlike gravitons or other packets of energy. These kernels of energy, like the shimmering light waves in Plate 9, might interact in a rich, dense webwork that gives space and time the appearance of being a smooth, continuous background stage. What has become of the gravitational field? At the very least, it has been replaced by a swarming population of virtual gravitons whose stringlike network structure creates space-time and seems to confer upon the gravitational field a measure of graininess. Because each graviton supposedly spawns bazillions of other gravitons in an orgy of self-interaction, physical space-time is populated by a continuum of collision events between one graviton and another. This makes gravity's spiderweb over the Abyss very finely meshed, especially at the Planck scale, so that every conceivable division of space leads to a nexus of gravitons that supply an even finer carpet of underlying coordinates in time and space. And this spiderweb may exist in more than four dimensions. The information concealed in the shapes of the other hidden dimensions could, like some cosmic marionette, invisibly regulate the entire gamut of behavior that we see around us in the physical universe.

There is hidden crucial information about the world in the quantum vacuum of the virtual world, and there is missing information about what distinguishes one particle from another, in the four-dimensional world where all things seem to operate and to which we have direct observational access. The "bumps in the night" that we vaguely sense are the invisible confederates that must work their hidden magic upon the world in order for reality itself to exist. These invisible confederates are hidden archetypes for all of the fundamental particles we experience in the physical world in both its virtual and measurable parts. As pieces of the gravitational field, they exist behind the scenes. They force us to live in a three-dimensional world because the particles and fields out of which we are constructed exist only with a large extension along these three dimensions. We can't turn a corner and fall into the fifth dimension because this other dimension is not big enough to hold us. As we marvel at the new landscape of space and time, however, there is in all of this hidden lawfulness and activity a newfound mystery to confront.

Out of the infinitude of these virtual particles that come and go invisibly in the cosmic gravitational field, an almost infinitesimal minority have been promoted into real particles following the specifications set out by hidden patterns tucked away in the nooks and crannies of other dimensions of space-time. Some gravitons have been converted into real quarks and electrons, which remain, though they are held together in gravity's field by virtual networks of energy in which they are still imprisoned out of view. But why is it, given the richness of the hidden virtual world and its gravitons, that all of Nature isn't perpetually trapped in this invisible world? Compared to the nearly infinite numbers of things "going bump in the night" within the gravitational field, even all of the elementary particles that make up the stars and galaxies of our world are insignificant trifles, a mere thimbleful of sand grains among all the world's beaches. How did the vacuum state of virtual particles lurking in the Void of the gravitational field offer up the physical world? To explore this question, we have to understand how the Void stores and releases energy.

9

A FIRE IN THE WHOLE
The Unstable Void

I realized that what I embody, the principle of life, cannot be destroyed. It is written into the cosmic code, the order of the universe. As I continued to fall in the dark void, embraced by the vault of the heavens, I sang to the beauty of the stars and made peace with the darkness.

—Heinz Pagels, *The Cosmic Code*

On a cloudy Saturday in June 1966, when I was a teenager, my family and I were visiting relatives in Sweden. As a special treat we decided to pay a visit to my cousin Dan and his family, at their summer home in Timmervik. It was a typical cloudy day, warm enough to be outdoors either at their beach or exploring the marshlands. Dan and I decided to make a short trip into the wilderness on his motor scooter. Setting off on the dirt road that passed their house, we rode for two miles and stopped in a spot where all we could see was a treeless, rocky marshland with boulders strewn all around the landscape, perhaps left over from the thousands of feet of ice that once covered this region. There were seagulls flying high up in the gray-cast sky, calling to each other. Their cries echoed through the uninhabited tracts of land, making it a rather melancholy scene. We immediately decided that the neatest thing to do would be to turn over every rock we could find to see what was underneath. Would we find a snake? A

193

mole? Would we find one of those elusive Swedish trolls? I spied a large rock about five feet from the roadside. Other than its size, there was nothing to distinguish it from the thousands of other stones that littered the marsh. It was a chunk of smooth red granite, a very common feature of the Swedish countryside, the bane of generations of farmers who had to extract them from their fields. I walked over to it, gave it a shove, but it was a bit too much for even an overexcited thirteen-year-old to move. Dan came over, and together we pushed the rock aside. We could tell it hadn't been moved in a very long time because the ground beneath it fit the contours of the stone exactly. We expected to find just a barren expanse of dirt, perhaps a few ants, but that's not what came into view. We both stood there staring. There, under the rock, was a five-krona bill, a piece of Swedish paper money with King Vasa sternly looking back at us. What was it doing there? What were the odds of our finding it?

When I was ten years old and had just finished mowing the lawn, I decided to get my beach blanket and go out in the backyard to lie in the sun and work on a nice California tan. It was a perfect summer's day, 85° Fahrenheit. I frequently enjoyed just lying in the backyard listening to the various urban birds and smelling the aroma of new-cut grass and the many flowers my mom had planted everywhere. She had created a little nook in the yard with a small pond surrounded by dark green, frondy plants. It even had a waterfall created by an electric pump. If you wanted to hear a babbling "brook," all you had to do was throw a switch. As I was lying there watching the clouds, thinking the long-forgotten thoughts of a ten-year-old, I happened to glance at the pond, and there was a turtle. Not a plastic one to add realism to the scene, but a real live turtle. In Oakland, California, in the middle of an urban neighborhood, you don't expect to see a wild turtle in your backyard.

Nature is full of surprises, but when the surprise gets too bizarre for normal logic, we reach for other explanations, whether it's for a small turtle materializing in an urban backyard or for Swedish money under a rock in the middle of nowhere. We rightfully assume someo person was responsible rather than a garden elf or a troll. Perhaps the turtle was someone's pet that had been let loose. Perhaps, just for fun, my cousin had played a prank on me by planting the money and then "randomly" driving me there to uncover it. It wasn't really my idea to stop at that spot.

When little things seem out of place, we rightly suspect that a human was responsible. When big things seem wrong, we take a different viewpoint. For thousands of years, when impressive earthquakes or droughts descended on us, we blamed Nature or some unkind deity. When it comes to the world at large, we never really believe that we can have any impact on it. We conduct atmosphere-altering experiments to increase its carbon dioxide levels and decrease its ozone. We see the beginnings of global warming and waste countless years embroiled in controversy over natural versus man-made causes. We destroy thousands of species of life forms on this planet every year. Then when someone tells us we are in the midst of the Sixth Extinction through our own thoughtlessness, we just laugh. Surely our destruction of the Amazon rain forest doesn't really matter, and besides, who really cares that some nameless beetle species succumbed? Scientists can paint you many stories of Creation that could have happened. They can also tell you some interesting stories about doomsdays that could have occurred, or may yet. Sometimes the line between the probable and the improbable can become very thin.

Before the first atom bomb was exploded, some scientists believed there was a slight chance the explosion could create a catastrophe that would destroy the atmosphere. It sounded like a crazy idea, but the worry wasn't completely without merit. The concern was based on the idea that the fission process in the bomb, once unshielded, would stimulate a nuclear reaction in the atmosphere. Like an expanding ring in a pond from a tossed stone, wider and wider regions of the atmosphere would become engulfed by this inferno. In a matter of a few hours, the Earth's atmosphere would be destroyed. Life would end. It sounded like a completely plausible process, given what people knew at the time. There was only one stumbling block: A calculation done by Manhattan Project scientists who seriously considered this possibility showed that the nuclear reactions could not run away like that, because the atmosphere was just not dense enough. It was true the atmosphere would heat up as hot as the Sun right near the blast, but the devastation would not spread around the world. The concern quickly vanished from all but a few historical accounts. Even there, it only served as a minor footnote to frontier science. In some accounts, the concern actually took on the cast of a "loony idea" that scientists looked into just for the fun of it. But what if Nature had been unkind and the calculation had been less definitive? Would we have still gone ahead with the blast?

Since then, science has not concerned itself with major global impacts other than laboratory "bugs" that might accidentally get out of the lab and cause a worldwide epidemic. Many people feared that genetically engineered viruses designed for innocent research purposes might mutate into Frankenstein monsters and become suddenly lethal. These are real concerns that even genetic researchers worry about, and the government meticulously regulates such laboratory research. In another sphere, we recently had our fill of the Y2K computer bug predicted to turn computers topsy-turvy worldwide. But after billions of dollars were invested in rewriting computer codes, the calamity never happened, and on January 5, 2000, many people claimed it had all been a big ruse. A year later, the media continued this same story again, never mentioning the billions that had been invested since 1998 to ensure exactly that safe outcome.

Astronomers have finally gotten the support they need to identify all of the asteroids that cross Earth's orbit, especially the nasty ones that might someday collide with Earth and cause some very real damage, even our own extinction. This isn't a joke. The geological record is clear enough about how often Earth is hit by large bodies. Every hundred million years or so, this planet gets whacked by asteroids five to ten miles across that are big enough to cause mass extinctions. The last one was 65 million years ago, and the dinosaurs paid a fatal price. Every few hundred thousand years, Earth gets pummeled by rocks several hundred yards across like the one that gouged out a meteor crater in Arizona. Every century or so, an object many tens of yards across hits Earth. The last of these occurrences was in 1909 in Tunguska, Russia. Finally, every year or so, average people report having seen a huge fireball streak across the sky. The biggest of these objects can reach several feet across. The U.S. military has a worldwide net of acoustic sensors that listen for infrasound pulses. About every year, they hear the atmospheric detonations of incoming bodies that explode with the power of several thousand tons of TNT. Earth also gets pelted by millions of meteors every year in a steady interplanetary rainfall. Occasionally people, houses, and cars are hit by meteorites, but these are infrequent events, and the meteorites capable of destroying entire cities are even rarer.

Catastrophe forecasting is also a concern of physicists, but the things they worry about border on science fiction. Although we understand that the Void exists as a physical agent in the universe, we don't really understand enough about it to know how it works. Our ignorance means we can imagine many ways it could misbehave. There is as much a chance for it to be totally benign as there is for it to act in a way that is not only unexpected but undesirable. In the most sophisticated theories we have, the Void can take on many different forms, some of which could exhibit Jekyll-and-Hyde behavior under certain circumstances.

In its most benign form, the Void is simply a passive medium physicists can stimulate to produce matter. Once the matter is extracted, there is no lasting effect. The Void heals itself and resumes its mysterious role. Up until the 1970s, this was the common view among physicists, at least to the extent that they even thought about it at all. Then a new version of the Void started to become known, offered in a theory claiming to unify the electromagnetic and weak forces. The Void—the physical vacuum—contained Higgs fields, which interacted with each other and gave the Void latent vacuum energy. This is very much like a magnetic field in a bar causing the bar to have latent energy because of the magnetic stresses. What was more curious about this theory was that depending on how vigorously you slammed bits of matter together, the vacuum could have two different energies. Both of these energies were the lowest possible ones that the Void could have. They were different in much the same way that a North pole is different from a South pole on a magnet. Now the question became: In which of these "lowest energies" was our vacuum located? It's a strange feeling to look out at the Void in the night sky and imagine that there could be another companion to our own tucked away in some corner of the world, just as matter has an antimatter partner. It's even scarier to think that this other Void is right here under our noses, and if we were to unbalance some invisible pencil balanced on its point, that might send us rushing down the energy ladder into some deathly doom.

Imagine living in a bar magnet. In your microscopic living room, you are reading the daily newspaper. You look up at the compass on the wall, which always reads "North." According to your parents, your grandparents, and the history of your species, it always has. Then one day, your world starts to heat up. No one knows why. As you watch the compass, it starts acting strangely. It begins wagging back and forth aimlessly. By this time, the temperature has reached a sizzling 800° Fahrenheit. Then, as mysteriously as it increased, the temperature begins to decrease. As you watch the compass, you can see that once again magnetism has returned to your world. The compass settles back down, but now, in a cosmic miracle affecting every living thing in your world, the compass says "South." All of a sudden, things that you had designed in your world that took advantage of its North direction now stop working. Everything has to be retooled to accommodate the South polarity.

Physicists exploring electroweak theory, and its big brother grand unification theory, have had to drag along in their mathematics a jumble of features that paint a very bizarre picture of the Void. It seemed that each time they unified one force with the others, they had to include a new energy level in the Void where such unification could take place. Every field they are forced to add to separate these forces in a cold universe puts its own invisible stresses on the Void, filling it up with latent energy like water pooling behind a dam. Taken as a whole, think of the Void as a mountain range. At the summit of the highest peak, there is the high-energy world of perfect unification. All three forces become indistinguishable. If you were to take a swan dive off this peak, you would hit ground with an awful lot of energy. As you walk the very long trail down from this summit, you come to a hanging valley where this unity is lost, so the strong force looks different from the weak and electromagnetic ones, which are themselves still unified. To get out of this valley, you have to climb up and over another smaller hill. Again you are on the down slope that leads you to a flatland area where all three forces now look different.

Different theories put the hanging valley at different elevations. Some even add a few more valleys to negotiate as you cool down the universe. Most of these valleys are easy to spot because you only enter and leave them if you can get things hot enough. The closest of these valleys is believed to be the one that allows electromagnetism and weak forces to unify. It's way up there at an energy of 300 billion volts, but until you can hike up to that valley from our much colder universe down here in the flatlands, you really don't know what to expect. Some theories, such as string theory, say that there may be other valleys, too. There are rumored to be some mighty peculiar animals there as well. If it does turn out that physicists discover the Higgs particle in the next few years, we will have to get used to the idea that we live in a most bizarre world with a complex Void and mysterious hanging valleys. It is from this world that a whole new round of ghost stories emerges.

There are some physicists who worry that reaching one of these high valleys even for the briefest moment may inadvertently unleash an avalanche that could cause major destruction. Like the Manhattan Project scientists worrying about incinerating the atmosphere, a new generation of physicists has been concerned about what could happen if their experiments went terribly wrong. Not just the Earth's biosphere, but the continuance of the entire universe as we know it might hang in the balance.

At a temperature of 1,000 trillion degrees Fahrenheit, most physicists expect that the Void will switch over to a new state where the electromagnetic and weak forces begin to look identical. Also, particles such as electrons and quarks will lose their mass. We have to take these ideas seriously. Physicists have been astonishingly accurate in predicting the discovery of new particles over the years. Between 2002 and 2007, this unexplored physics will be investigated by a new generation of powerful accelerator laboratories such as Fermilab in Illinois and CERN in Geneva, Switzerland. Physicists will accelerate electrons, protons, and their antiparticles and collide them together.

For less than a billionth of a second, they hope to create a minuscule spot in space that has a temperature of 1,000 trillion degrees Fahrenheit. It will be like a blowtorch of pure energy acting on the frozen ice of the Void. What will happen? Physicists don't really know for sure. Over the years, they have convinced themselves the effects will probably be purely local. However, there are some unpleasant possibilities that have been investigated over the years.

Suppose that the Void we live in isn't really the one with the lowest energy? We are pretty convinced that since the big bang the universe has rolled down the Olympian mountain of energy and has safely passed through many possible valleys, including the hanging valley of the electroweak force. It has continued to roll down to the very chilly flatland world where we live now. All of this excitement took place during the first microsecond after the big bang. We look around and see the frigid cosmic fireball radiation at 2.7° above Absolute Zero. It really looks like the universe has gone downhill just about as far as it can. But what if right now, we were instead caught in yet another hanging valley? Suppose that there was an even lower-energy world. Our universe has a temperature of only 2.7° above Absolute Zero relative to the current zero-energy of the Void. But suppose this "zero" was perched like a roller-coaster car on the top of a high track, located 1 billion degrees above the bottom of the railway? What would happen if we accidentally did something in our laboratories that disturbed the current balance, triggering the final descent? There are three logical possibilities according to the equations: Physicists might simply create a piece of space with momentarily exotic conditions that will vanish as soon as the collision is over; they might create a region of exotic physics that will persist but not grow because of the action of some new unexpected process; or they might create a region that will grow in size, converting some or all of our universe into a new phase of matter and energy.

Back in 1976, Peter Frampton at UCLA looked into the mathematical side of this issue. He found the first hint that something spectacular could happen if the universe was built in just the wrong way. By this time, many theories required that the Void must contain several different energy levels, but Frampton found that if the levels were too close together, a very unpleasant condition could be created. Just as neutrons and protons can escape from inside the nucleus of an atom, the whole universe could tunnel like a mole through the energy barrier separating adjacent Voids. The other thing that emerged from his calculation was that like the compass in the bar-magnet world, the mass of the hypothetical Higgs particle could tell us just how close we were to this doomsday scenario. A quick calculation published in the November 22, 1976, *Physics Review Letter* told the story. If the Higgs particle was heavier than about five times the mass of a proton, there wouldn't be any problem. The energy level of the Void we are in would be perfectly stable. But if the mass was less than this, the Void would be unstable; we would be in deep trouble and, in a sense, living on borrowed time. At any time, a bubble of the lower-energy vacuum could spontaneously form in some corner of our universe. Expanding at the speed of light, it would eventually overtake all the stars and galaxies in an ever-growing bubble of devastation. A new universe would literally be born out of the ashes of an older one—ours.

Frampton's 1976 paper, "Vacuum Instability and Higgs Scalar Mass," was followed the next year by "Fate of the False Vacuum," by Sidney Coleman at Harvard University. Coleman's more elaborate calculations of exactly how the current Void would be replaced by a new "true vacuum" were even more distressing. Coleman's paper had already set the stage for these disturbing possibilities by investigating how bubbles of true vacuum (our stable kind) form within an arena of false vacuum, the kind of vacuum caught in one of those hanging valleys. He borrowed heavily from the simple physics operating when bubbles of steam form in water near its boiling point. If the Void behaved similarly to boiling water, we would be in for a very spectacular and incinerating ride, like the crashing of the bridge in Figure 4.2 into some unimaginable abyss.

Coleman discovered after considerable mathematical study that if a bubble is too small, it simply collapses back into nothingness. This happy situation would be completely overturned if its size at formation was just right. It would continue to expand at nearly the speed of light until the false vacuum we currently live in had been converted over to the true vacuum throughout the universe. This conversion would release a huge store of latent energy in the Void. If this decay rate was only a few years, then you end up with a second big bang after the "big one" that originally started the universe out in the false vacuum phase. But Coleman also added, "If this time is on the order of a billion years, we have occasion for anxiety." If a piece of our universe were suddenly to erupt into a true vacuum bubble, we would never know it. Its outer wall would expand at nearly the speed of light. We would have no warning of its approach until it was too late, and we certainly could not get out of the way of this cataclysm, which would be incinerating all of space in its wake. If the initial state of the vacuum contains some particles embedded in the false vacuum, what happens when a bubble wall encounters those particles? What, exactly, could happen to us? By 1977, no one knew the answer, and the prospects seemed very bleak. Sidney Coleman and Frank De Luccia published a paper in 1980 in which they described what might happen to the universe if it had ended up in the wrong kind of vacuum at the big bang.

> The possibility that we are living in a false vacuum has never been a cheery one to contemplate. Vacuum decay is the ultimate cosmological catastrophe; in a new vacuum there are new constants of nature; after vacuum decay, not only life as we know it is impossible, so is chemistry as we know it. However, one could always draw stoic comfort from the possibility that perhaps in the course of time the new vacuum would sustain, if not life as we know it, at least some structures capable of knowing joy. This possibility has now been eliminated.

The bottom line in all these speculations is that a very small part of the Void might go "up and over" some energy barrier, which for now prevents our universe from resuming its roll down the slope to this even-lower energy for the Void. If physicists accidentally triggered this cataclysm, particles would collide in our machines and energy would disappear into the Void to temporarily melt it. It would then cool back down and trigger an expanding bubble of the new Void that would fly out from this nuclear Ground Zero at the speed of light. In the wake of the wall of this bubble that would reach the Moon in two seconds would be a new Void. The physics of the old false-vacuum Void, together with its virtual particles and fields, would be unalterably changed, but so might such things as the mass of the electron and some of the other fundamental constants. Could our familiar world survive the aftermath? The transformation of the false vacuum to the true vacuum would be so fast we could in principle go from a normal world to nonexistence faster than our neurons could transmit physical changes to our consciousness. We would feel no physical pain. In a different scenario, if the change were very subtle, we would not even notice that anything had happened. Physicists, however, might discover that the mass of the electron had altered slightly, compared to what their notes and handbooks had said it was prior to the changeover. In between these two extremes would pass a bewildering set of alternate possibilities, some filled with dreadful pain and inconceivable despair.

A few years later, Alan Guth and André Linde developed "inflationary cosmology," and this found a permanent place for the false vacuum–true vacuum changeover. It was relegated to a time near the origin of the universe. Although the changeover would be devastating today, back then it had a very positive impact on the universe. During the time the universe was in the false vacuum, a tremendous energy built up in space that caused it to double in size every trillion-trillion trillionths of a second. It came to an end a few thousand doublings later, when the false vacuum changed over to what was thought to be the current true vacuum, perhaps in the way that Sidney Coleman had imagined several years earlier. Everything seemed to be okay, at least for now, according to this train of thought. But then in the 1990s, astronomers began to uncover several lines of evidence that pointed to a disturbing fact. The universe was not expanding the way it ought to if it really was in a true vacuum state. Today, we look out in the universe and we see its expansion accelerating, just the way we would expect if we were living within a Void that was in a false vacuum state with a small amount of residual energy. The calculations show that this latent energy in the false vacuum acts like an antigravity pressure. If the Void changed over to a true vacuum, this pressure would vanish and the accelerated expansion would come to an end. A phase of constant expansion would take over. Therefore, we are already living at a time when something very peculiar is going on in the Void. If this really was the case, then the Frampton-Coleman doomsday scenario might still be in the cards. What could push the universe over the edge? Could physicists accidentally do that?

The doomsday scenario found its way into the physics journals in the late 1970s, but it was quickly put out of business once physicists looked carefully at reactions in Nature that could also have triggered this changeover in a much less artificial way. The focus turned toward those pesky gnats of the universe called cosmic rays. The formation of the bubble in our universe is one of those things that, as Peter Frampton suggested in 1976, could be induced by the collision between two particles that could feed enough energy to the Void in the false vacuum state to bring on a bubble. Cosmic rays carry a lot of energy. Astronomers have studied them since the 1940s and over the years have built up a pretty sound understanding of just how high their energies go. The most powerful ones have energies up to 10 trillion volts. These particles, mostly electrons and protons, collide from time to time with interstellar matter and the gases in the atmospheres of planets like Earth. A few hundred of these enter Earth's atmosphere every second, triggering air showers of subatomic particles that reach the ground. In the midst of all this opportunity for Nature to unleash the collapse of the Void, we are still here. Nature has performed this experiment of melting the Void countless times across the universe. The good news is that nothing has ever happened. Physicists were pretty sure that the technical arguments against the doomsday scenario were in good shape, and this exercise actually taught them a lot about the Void. It is very robust and hard to unhinge. Even Frampton's estimate of how heavy the Higgs particle would have to be to signal an unstable vacuum seemed to be comfortably exceeded by a number of very aggressive searches for the Higgs. The Higgs particle has to be heavier than about eighty times the mass of a proton to have completely avoided detection so far, over ten times the critical mass in Frampton's original analysis. An interesting final caveat to this is that what the universe does here and there with a couple of collisions from time to time will be nothing like what physicists will soon be doing in the laboratory. The billions of collisions per second that the next generation of accelerators will be creating are of a different order of violence to the Void. We may not be entirely out of the woods yet.

I wrote an article for *Astronomy* magazine in 1985 titled "The Decay of the False Vacuum" and tried to spell out what the implications of the vacuum decay process were for cosmology. All of my discussions about artificially causing this event today wound up on the cutting-room floor. Instead, I spent my time describing the new inflationary cosmology and the great things it was doing for big bang cosmology at the time. There have only been a few other popular articles written about this since then. Every one of them has also adopted my strategy of focusing the public attention on the very distant "inflationary era." Vacuum decay was seen as a good thing for the infant cosmos and not as something that could happen today. It was a fair assessment, given that by the mid-1980s even seasoned theoreticians had concluded that we were completely safe from any new transformations. One group, however, did hold out the possibility that trillions of years from now, the Void could make a final plunge and spawn a new daughter universe from its own fabric.

Edgard Gunzig, Jules Geheniau, and Ilya Prigogone were among the vanguard of physicists in the 1980s who were pushing the true vacuum–false vacuum model as far as it could go. In 1987, they landed on a rather startling new idea. Inflationary cosmologists had focused on what would happen to the changeover if it occurred when the universe was very young. Could it also occur in a natural way when the universe was very old? It looks that way. When this new transformation happens, the result will be spectacular. As our universe grows more uniform in the far future, it might actually become the "mother space-time" for an entirely new universe literally torn from its own fabric. Once born out of this future Void, this child universe would become totally disconnected from our universe so that no direct spatial or physical link remains to us. Only a black hole would likely remain as a mark of the aneurysm that had formed. In time, even this black hole would evaporate, leaving no trace or cosmic afterbirth. By this remarkable process, a universe may actually be self-creating and eternal. This creation of a new universe out of the fabric of an older one need not have happened just once. It's just as easy to create an infinite number of child universes as it is to create only one, once the mother universe becomes old enough. In fact, what we know about the Void almost seems to demand this possibility.

So far, these explorations of the Void have been purely theoretical, with little connection to the world we know either by experiment or everyday experience. They remain buried in the technical literature. The public had never been subjected to the rapidly changing landscape of the academic debate. But in 1995, this all changed. An idle comment by a physicist was carried by the news media, triggering a public confrontation over the nature of the Void and whether bad things could really happen. No one knew whether an experiment could by accident push us over the brink to trigger an ecological catastrophe, but some people were convinced they had to take a stand. Fermilab was picketed by Paul Dixon, a psychologist from the University of Hawaii, and his supporters. They were justifiably afraid that the new Tevatron accelerator would be the home of the next supernova in this corner of the Milky Way.

A spectacular day of triumph for Brookhaven National Laboratories was anticipated for June 2000, and it certainly needed some good news. In prior years, Brookhaven had been dragged into court over the health effects of beams of neutrino particles that were spewing out across nearby towns on Long Island. This time, Brookhaven had built a new tool called the Relativistic Heavy Ion Collider (RHIC) and was anxious to fire it up as scheduled in what would be a much-awaited, exciting moment for physicists. The $365-million machine was supposed to collide the nuclei of atoms such as gold together at more than 99.99 percent the speed of light. In the maelstrom of colliding quarks and gluons that would blaze for a few billionths of a second, physicists hoped to melt the vacuum and see what would result when it refroze. This experiment of vacuum modification would not cause an accidental change in the Void like a cosmic ray. It would be a detailed, premeditated probe of the Void to tease out just how it worked. Physicists planned to create an entirely new state of matter never before seen in this universe since the time of the big bang: quark-gluon plasma. Some scientists who went on record to describe this experiment used the standard, laconic language of physics to express their opinion about what might happen. Mark Alford, a physicist at MIT, commented to Peter Weiss, a *Science News* reporter, "If we find out when we heat up the vacuum that the properties change dramatically in some way—like water changing into steam when we boil it—then that would be very interesting." If everything worked as planned, the colliding gold atoms would compress the constituent particles to a density 100 times higher than normal nuclear matter, spiking a massive temperature of over 1 trillion degrees. Some of this energy would actually change more than the properties of the quarks and gluons. Physicists hoped that it would also burn a hole in the Void, raising its temperature high enough to alter its properties.

Overlaid on the Void and occupied by the virtual particles of electromagnetism and the weak forces is a second vacuum that is a product of how the strong force operates. In this layer of the Void, quark-antiquark pairs fill up so much of the Void that they become locked together into what physicists call a "chiral condensate." This condensate, like the atoms in a bar magnet, exerts a force that acts over a very short distance that is in fact smaller than the nucleus of an atom. Physicists hoped that this vacuum force, like the Casimir effect in QED, might be the missing link that causes quarks to be confined inside nuclei. Certain theories also suggested that this Void seems to force gluons to flow inside tubular structures. By melting the Void, RHIC would let physicists study how the Void works and help them decide which one of many models for the Void was on the mark. By June 1999, the accelerator was nearly complete. The time was soon coming when the first gold atoms would begin to circulate around the 2.5-mile racetrack. The first controlled attempt to actually manipulate the Void at very high energies was nearly at hand. But then other concerns began to intrude.

Madhusree Mukerjee, a staff writer for *Scientific American*, had written a short article, "A Little Big Bang," in the March 1999 issue of the magazine, explaining with enthusiasm how physicists were about to create conditions not seen in the universe since the big bang. Following the article was a letter by Walter Wagner, a reader of the magazine from Hawaii, asking whether scientists knew for sure that the experiment was safe and would not create a black hole. Frank Wilczek, a physicist from Princeton University, responded that it was very improbable that the Brookhaven machine would create a black hole, but in one of those casual comments that theorists love to deliver almost with a smirk, he said that it might instead create "strangelet" particles that could swallow normal matter. This exchange prompted the Sunday *New York Times* to publish an article on July 18, 1999, with the provocative title, "Big Bang Machine Could Destroy Earth." Wagner then filed a lawsuit to prevent the RHIC from being switched on. Children wrote impassioned letters pleading with Brookhaven scientists not to destroy the world. Even the downing of John Kennedy Jr.'s plane was eventually attributed to Brookhaven's having created a black hole that had swallowed him up.

Soon after the experiment was begun as planned, Robert Crease wrote in the July 2000 *Physics Today* magazine that there were many lessons to learn from this event, not the least of which concerned how quickly technical information can find its way into the media. Scientists' traditional uncertainty could all too easily be turned against them because the rest of us want to "know for certain." Both Brookhaven and CERN put together panels of eminent scientists to review risks of this sort more thoroughly. All they could turn up was a set of very improbable events, a fact that loomed large in the arguments against catastrophe. The impression prevailed that nature had already conducted these experiments through cosmic ray collisions. However, the disturbing fact was that the calculations were only done *after* the machine was about to be turned on. Before the machine was built, no one had bothered to look into what might have been a devastating environmental crisis, to say the least. Again, doomsday scenarios triggered accidentally by physicists were not the sort of thing that anyone took seriously, for the same reason that no one seriously imagines that any human action can create a hurricane or an earthquake, let alone destroy Earth and the entire universe. Very few science fiction stories even bother with such an improbable possibility. Besides, wouldn't you like to see what happens when you torch the Void?

Wagner's suit failed when Brookhaven was acquitted in 1999, and the experiment went on without a hitch. But we may not be entirely out of the woods yet. In the next few years, an even more dramatic experiment will begin to probe the Void. There are already some new theories queuing up to predict what might be discovered. M-theory has spawned a new theory for gravity in which the additional dimensions are not vanishingly small. They may actually be seen under tabletop laboratory conditions. These macroscopic additional dimensions (MAD) will make themselves felt at distances of just under a millimeter if we live in a five-brane world and about the diameter of an atom for a six-brane world. The more spectacular prediction by MAD scientists is that for five-branes, gravity will become as strong as the other three forces at atomic scales. If this is the case, MAD scientists will start to see the effects of quantum gravity within the next generation of accelerators. Could one of the by-products be the creation of quantum black holes? Here we go again.

In previous versions of quantum gravity, space-time turns into a frothy sea of forming and evaporating black holes, each of them carrying about 0.00001 grams, with a size of a billion-trillion trillionths of an inch and destined to live scarcely 10 billion-trillion-trillion trillionths of a second before evaporating back into the Void. Also at this scale of sizes and energies, all of the four forces become unified in some kind of stringy world. If MAD scientists are correct, this extrapolation is all wrong. Using their "braney" science, at about one millimeter or less, the four forces could all be unified at the distances probed by energies near 1,000 GeV in the next generation of accelerators. If this is the case, then it may also be true that the quantum graininess of space-time will start to appear at these energies. Quantum black holes might start to form. Also, as more energy is poured into these quantum black holes, they would grow in size. Could these artificial black holes do any damage? Could one of them fall out of the accelerator and consume the Earth? Again, this particular doomsday scenario seems very unlikely.

We cannot know exactly how black holes really work without direct experimental confirmation and study. Since all of their properties are deduced from mathematical study, we have only an untested theory of how they might work to serve as a guide. One of the most interesting theoretical discoveries made about black holes over the years has been that they probably evaporate because of quantum mechanical processes. The intense gravitational forces literally rip virtual particles out of the Void and inject them into our world. From the outside, the process looks as though the black hole is actually emitting matter like a leaky bag. Small black holes with the most intense fields evaporate the fastest. Very big black holes with weaker fields take a much longer time to evaporate. The calculations suggest that if you were to create a black hole that had a mass of one gram, it would evaporate within a few thousand-trillion trillionths of a second. On a larger scale, a black hole with as much mass as a small mountain would last as long as the age of the universe to date. To make a one-gram black hole, you would need the output from a small atomic bomb focused on a region of space smaller than the nucleus of an atom. This is not within the engineering capability of any machine that is likely to be built in the foreseeable future. Therefore, our having to confront the aftermath of accidentally creating these planet-gobbling black holes gets pushed into the dim far future. The new theories might let us create them, but they will evaporate completely before they have had the chance to move less than the diameter of the nucleus of an atom. There is also an easier way to eliminate this whole scenario— just discredit the MAD theory that purports to make them in the first place.

Many groups of experimenters are even now conducting the painstakingly difficult tabletop experiments to check whether gravity starts acting differently at the one-millimeter scale. If they find nothing, then MAD scientists will add one more dimension to the brane we are hypothetically living in and explore the possibilities of a six-brane world. This will push the search for new gravity effects to about the size of a uranium atom. Given all of the other interatomic forces at work, the technical difficulties in checking this scale of gravitational effects becomes so difficult that this will keep experimenters busy at least until the new LEP II accelerator at CERN is turned on in 2005. But by that time, the threshold energy location of the new gravity effects will have moved up a notch well beyond where LEP II can probe. The critical time will arrive sometime between 2002 and 2004, when tabletop experiments will be underway exploring the physics needed to create artificial quantum black holes, but before accelerators can actually begin to probe the MAD energy scales directly.

It has taken us more than 5,000 years to reach the point in science where we can begin to see patterns in the Void and address them. The last hundred years have taken us on a particularly stormy journey far into cosmic space and deep into the nature of matter and field. As we enter the twenty-first century, investigators appear poised to discover the elusive Higgs particle, and perhaps even new dimensions to space-time. If the Higgs is not seen, physicists will need to reconsider over thirty-five years of research based on the Higgs process. In that case, the search will have been one of the biggest intellectual detours of talent and resources in all of scientific history, though it will still have been necessary to pursue a promising theory to the last decimal point before putting it aside. With it will also fall the current notion of a complex Void with multiple energy levels patterned after the similar levels seen within the atom and the nucleus. We will all breathe a sigh of relief that there is not some lower-energy abyss lurking in the Void into which our universe could fall. Perhaps much of string theory and M-theory will also be discredited. If the Brookhaven RHIC is unable to uncover traces of the mysterious but necessary quark-gluon plasma, our whole notion of a Void containing a patina of quark-antiquark pairs in a chiral condensate will have to be severely modified or eliminated altogether. If the accelerated expansion of the universe cannot stand the test of time as more data become available, then the cosmic Void is indeed a bleak plenum with only the cosmic gravitational field surviving from a more turbulent age. Finally, if MAD scientists cannot detect traces of large extra dimensions, the entire avenue of string theory will fall, leaving only the very much harder Planck-scale physics to deal with. It will also reinstate a Void much less willing to divulge more of its secrets to the eyes of prying scientists.

The odds that all of these scenarios will materialize seems very remote from my perspective in the year 2002. The quark-gluon state was seen briefly in several experiments in 1998–2000, so the Void will probably gain new bedfellows from QCD. Even the accelerated expansion of space has been independently confirmed by NASA's COBE satellite and recent balloon-borne telescopic studies of cosmic background radiation. Only the multidimensional predictions of string theory seem at odds with anything seen so far, although physicists may have spotted traces of supersymmetry, a critical feature of string theory that can be tested at low energy. So at least for now, we can only look at the Void with a trace of urgency and hope that the ancient mariner's map notation, "There be dragons there," does not apply. Although the Void seems a quiet place today, it is almost certain that it wasn't so in the distant past. It is there in the ancient past that we find clues to the origin of the material world emerging from the invisible shadow world of the Void itself.

10

THE SUNDERED WORLDS
The Creation of the Universe

Nothing makes the darkness go, like the light.
—Madonna, "Nothing Really Matters"

I was asleep in my bed after a hard day of digging out old stumps in our backyard. My family had just moved to a new home. The backyard was a real mess, with dead trees and ivy everywhere, which all had to go so we could plant a proper garden in the spring. I hardly remember the usual round of "good nights" I exchanged with my wife and two young daughters that evening. They told me the next day my head had barely touched the pillow before I was "gone." The exhausted, intense sleep I was to get that evening rocketed me into another world I have only visited a handful of times in my life—"waking dreams," they call them. I awoke to a semidark room I clearly recognized as our family bedroom. My wife and children were sleeping peacefully. My vision seemed a bit clouded, as though I were looking out at the world through a gauzelike shroud. There were no obvious sounds from the air-conditioning system, nor did I hear the newly familiar creaking noises of our new house. The most striking thing, however, was that I was completely immobilized. I couldn't so much as wiggle a toe or make a sound. And in the next few instants, I dearly wanted to utter a sound. There, at the foot of our bed, were three small creatures with odd-shaped heads. They seemed to have

213

entered the darkened room through the bedroom window, which I was sure had been locked securely. One of them stopped what it was doing, realizing that I was watching it, and hopped on the bed. Surely the motion would wake my wife. For some unexplainable reason, it didn't. The last thing I recall was it thrusting its hideous face within inches of mine. I woke up for a second time. The familiar sounds and aromas of our bedroom flooded into my senses, but my heart was racing in panic. I looked around the room warily, listening for the sounds of these awful intruders, but the world had again returned to the one I had always known. I didn't bother to check the basement.

Dreams seem real for a time, but the longer they play themselves out, the more they deteriorate into a cloud of nonsense. Our brain is constantly looking for things to do. Creating fabulous dreamscapes is one of its greatest and most mysterious talents. Even now, no one really knows why the brain has to dream. But it has become obvious after decades of research that if robbed of this creative outlet, the human psyche descends into the world of psychosis within only a handful of days. We also indulge in a much milder form of dreaming that lets us play with real-world experiences under strict mental control. Just as our brain needs to dream, it seems that we also need to tell stories and create theories to help sort out events in the waking world or just for the simple joy of entertaining ourselves. It is perhaps one of our greatest talents, at least after toolmaking.

Thousands of years of practice have made us pretty good at spinning tales—so good, in fact, that a well-crafted story can't be distinguished from a real experience, though some stories can be more hallucinogenic than many of our wildest dreams. The art of storytelling is in enticing listeners to suspend reason, to float along unfettered on every word as the story carries us deep into another world. Nothing can be more captivating than to hear a master storyteller spin a webwork of events into a fantastic tale of adventure, suspense, and mystery. Daytime stories range from pure fact to light-hearted fantasy. Without exception, dreams are much more imaginative. To keep us interested in daytime stories, a much stronger sense of obedience to the rules of cause and effect is needed than anything found in the ramshackle worlds we spin while asleep. Without an undercurrent of cohesiveness, without familiar hidden patterns and rules to count on, it's hard to suspend our disbelief. The story becomes threadbare, and our temporary flight from reality comes to an end. For other stories, by the time we arrive at the end, the story has in some sense only just begun. The final cadence is just a sustained chord drawing energy from each time the story is retold down through the centuries. These stories transcend the moment or the age. In their simplicity, they command our awe and deepest respect because the central characters are unfathomable, the main act overpowering.

The most ancient stories known to us are the Creation legends about how Earth was formed. Some are mere fables passed from generation to generation, mostly told to children, perhaps to quench the childhood habit of asking too many questions of busy adults. Other legends have a decidedly more serious tone. For millennia, high priests had direct communion with these stories. Eventually, they were written on papyrus or clay tablets so that no one would forget them or alter their meaning. No one knows who authored them. A committee? A single magnificent great thinker? All that is known is that the threads of this story bind many civilizations together in an unbreakable tapestry. When you consider how each thread winds its way through dozens of different cultures and religions, many of which have historically been at each other's throats for centuries, the similarity seems more than coincidental. The ancient Semites learned their Genesis story from even more ancient Babylon, Egypt, and still other civilizations before them without name or number. These stories begin not with a whimper and lofty phrases but with a powerful, emotional bang.

"In the beginning, there was darkness." This powerful thought echoes in countless Creation stories, from ancient Egypt to the modern tribes of the Zuni and Maori. There can be nothing more desolate or empty of meaning than an eternal darkness. Although the primordial state may have been without light, storytellers insist that it was not empty of content. All legends speak of an unimaginably mysterious substance or primordial water—"Nu," the Egyptians called it; to the Semites it was "the Abyss." It was the wellspring of all things. How did the ancient storytellers come upon the image of an origin for things buried in the depths of some dark, cosmic essence? It's easy to imagine tribal shamans leading an initiate into the kinds of caves our ancestors lived in 20,000 years ago or more. Down a passageway they solemnly walked, into the bowels of a mountain. Perhaps they carried a torch to light the way, perhaps they groped their way in total darkness. There in the dark, they heard the steady drip of water. As the torch passed along the walls, the shapes of animals suddenly appeared, their forms emerging from the darkness with the advance of the light—but always with the sound of water murmuring in the darkness just beyond the light's edge.

In our most ancient Creation stories we also have dark primordial oceans with hidden forms, waiting for light to bring them into being. This formless cosmic ocean enshrouded in darkness would have prevailed for all eternity were it not for a "mighty wind" or a "primordial spirit" or a "spirit of God" that swept over the waters like fire lighting a cave wall. No one tells us where this spirit or wind came from. Egyptian priests and Zuni shamans alike tell us it created itself out of the Void: "Before the beginning of the new-making there was nothing else whatsoever throughout the great space of the ages save everywhere black darkness in it and everywhere void desolation." The Old Testament says only: "Darkness covered the Abyss while [the] spirit of God, a mighty wind, swept over the waters." By all accounts, it is an ageless spirit, for the God of the Semites is a spirit of pure, absolute Being—Yahweh—though it is never offered that He created the waters themselves. We now have the raw material and a creative agent. All we need is something interesting to happen from their union. The storytellers don't disappoint us.

In story after story, the stage is set for the first wondrous act to take place. "God said, 'Let there be light.'" If an eternal, formless darkness seems too oppressive and fearful to contemplate, what can be more glorious or dramatic than replacing it with light? In four words, in a command uttered from the depths of the Abyss, light replaced darkness. All of eternity to come was completely transformed. The ancient Egyptian version of this event is a bit different from the Old Testament Genesis story. The Egyptians don't mention light as a significant element at all in their Creation story but take a different approach altogether, in which the phrase "his word awoke the world to life" plays a central role. In India, the world is said to vibrate with the echo of Om, a sound emitted at Creation, which still penetrates to the very core of existence itself. Across all of Egypt, throughout 3,000 years of history, they revered the Word. They wrote elaborate incantations to ensure safe passage into the underworld or to dispose of enemies. They obliterated names on statues to deny eternal life to unpopular pharaohs, because if no one knows your name, your spirit ceases to exist. And it was the sound of a word, spoken at a propitious moment, that could change the world or create a new one. Ra-Kheperu, the Creator and Bringer into Being, was alone in the Void, creating himself by uttering his own name. When Yahweh commanded light to come forth, it would be a far more powerful act in its simplicity. His first words banished the fear of the night, the darkness of the cave lurking in each of us. He replaced it with hope. This, according to Genesis, was Yahweh's first act. Then out of the waters, matter in all of its various forms came forth: the Earth, the sky, the heavens, and all the rest. Ra called them forth by naming them, because their forms were already in existence in his mind like figures on a cave wall waiting for the light to reveal them. Matter came forth from the waters, from Nu, from the Abyss.

In other traditions, in other lands, Creation legends celebrate the mystery of shapes and substance emerging from dark formlessness. In some stories, the world is capable of creating itself without divine intervention. The Maoris sing of light created from darkness, space created from spacelessness. Across the Pacific Ocean, Marquesas Islanders speak of a "primeval void that was swelling, whirling, and vaguely growing, a boiling maelstrom out of which emerged the foundations of space, light and matter." From India in the eighth century B.C., we hear how space condensed out of the Void, then how air condensed from space. The Hindu Rig-Veda tells how the gods were created only after the universe itself came into existence, but the gods themselves couldn't explain how they had come into being. The Hindus and Brahmans had ended the question of why things exist. If the gods couldn't answer, then surely mere humans would never be able to, either.

As century piled upon century and ancient stories solidified into accepted dogma, newer stories emerged from ancient Greek philosophical traditions, unrestricted by the mandates of mythology or religion. Philosophers were the new story makers at a time when Semitic, Asian, and Indian stories were already ancient. Like many traditional storytellers of the time, Greek philosophers believed the world had a beginning in time. It would one day come to an end in what the Vikings later called Ragnarok, or what the Old Testament called Armageddon. Many mythologies were very comfortable with a definite origin in time for the world. But the ancient Greeks were thinking about more than just the limits of religious knowledge. They were experimenting in a public way with the idea of logic. Unrelenting logic led them to explore Creation with a very different motivation than the kind that comes forth in any ancient writings. What the philosophers were trying to fashion wasn't really a story about Creation. It was an investigation of the details of the process itself, and it is here that the exploration took a fatal but unavoidable turn.

The new details offered by Greek logic somehow seemed to lack a soul, and lacked majesty and evocative imagery as well. In the end, the philosophers' logical comments on Creation as a process were not stories shaping a civilization; they were minor embellishments of no great personal consequence, mere opinions that held sway over the minds of a few individuals or their students. What did it really matter how God, or some other agent, actually brought the world into existence? What mattered was the relationship between humans, their gods, and the afterlife. Greek logic led to a series of revelations that had nothing to do with humans and gods or their interactions. Lucretius thought the guiding principle of his cosmology was that "nothing can ever be created by divine power out of nothing" and that "Nature resolves everything into its component atoms and can never reduce anything to nothing." Later on, Plato offered an even more basic view of the first state of the cosmos: "The matter of the visible, sensible world has a nature invisible and characterless, but partaking in some very puzzling way of the intelligible, and the very hard to apprehend." These may have been fascinating speculations driven by logic, but they were totally irrelevant to the traditional Creation stories of the time. Even new religions such as Christianity had no obvious interest in what Greek philosophers had to say about Genesis. Thus, without the support of religious traditions or inspiration by anyone's gods, Greek thinking about Creation fell into an intellectual black hole for thousands of years. If no deity could speak for these new insights, then neither would any mortal.

Even today, scientific exploration of the world is not specifically sanctioned by any religion at the level of the weekly sermon. Discoveries about the entire physical world are seen as incidental or at worst as an unacceptable challenge to the authority of religious teachings, rather than as evidence in support of it. After all, the purpose of the physical world is to be a temporary vessel, which will be subdued and discarded by humans at the moment of spiritual transcendence. The physical world is a curious anecdote to our existence, but not of itself worthy of much consideration when other, more pressing spiritual matters constantly take center stage. Yet, just as light and dark come together to create images like those in Figure 10.1, we also see that the scientific and religious descriptions of Creation are not the opposite sides of the same coin but complementary views that reinforce the essential mystery of existence.

Each of these ancient Creation stories is in many ways very different, but all are beautiful in their ageless simplicity. Compared to the enormous numbers of words that would come afterward to lay down the rules for human behavior, Creation stories are brief, to the point, and awe-inspiring. They celebrate the creation of life from nonlife, of existence from oblivion. These are the issues that occupy each of us and stir the emotions. We feel a kinship with the storytellers because they were people like ourselves, gifted with the same intelligence and curiosity about the world. Like us, they wondered how they had come to be. It is here that something truly puzzling begins to emerge. The stories they created to help them understand their origins tap a common kernel of knowledge transcending culture and age. By the time ancient editors finished revising each story into its recent form, the stories flowed into a startlingly common pattern. Each story began with a primordial emptiness, which had perhaps existed eternally, though certainly without light, and was probably very chaotic. It was

FIGURE 10.1
Patterns of light and dark are inter-twined throughout Nature and evoke a sense of awe and mystery when accompanied by the right story. Modern physics often forces us to accept competing ideas (particle-wave, matter-energy, time-space) as part of a complete experience of Nature, much as a visually ambiguous image forces us to experience conflicting ideas (old woman–young woman) as part of a single whole.

the elegant and simple counterpoint to a crowded world. Then, some ancient spirit embedded in the Void or perhaps apart from it in some unimaginable way acted on this primeval emptiness to bring into being the various elements of the physical world. Again, what could be simpler? A single powerful act to turn dark emptiness into a filled and vibrant universe. Among all the renditions crafted by storytellers, editors, and chroniclers, we see the same blinding light of insight penetrating through the shadows of confusion to a remarkably similar starting point, time after time, in civilization after civilization. It is easy to feel justified in thinking that this guiding principle was an inspiration from God.

What scientific insight can now bring to the Creation story after 5,000 years is a clean slate, a fresh language. The words are so new they are not even a part of our common languages, yet they were created by communal acts of understanding that go beyond a specific culture or language. Science has reached an amazing pinnacle in the last hundred years that far outweighs its three-century youthfulness in the eyes of other traditional ways of gaining knowledge. We have entered into a whole new style of exploring our universe, one in which detailed observations, along with inspired logical deduction, have become a part of the craft of creating the story. There is now a very keen and undeniable understanding of many of the basic ways in which Nature works, of how it has put matter together into a dazzling number of shapes and forms, from stars to the delicate aurora shown in Plate 10. This insight is improving every year: It is detailed, objective, experimentally verifiable, and it can be measured; it can generate spin-off technologies. All of this would not be possible were it not for Nature's being so obliging.

No one really knows why the universe is so rich in detail. The question may itself be its own answer, because without all of this richness, human observers would not be here to even worry about it. One of the spectacular things about the way Nature seems to be put together is that details eventually link up across a wide spectrum of sizes and systems, from the inner workings of atoms to the color of a rose or the gossamer rings of Saturn. You start in one small corner of the cosmic puzzle and ask a simple question, a question that even a child can pose. The answer takes you into other corners of the puzzle to which it is connected. Physicists studying how nuclei are put together eventually link up with astronomers who study how supernovae explode and produce the elements that make up our blood and bones. A question resisting a quick answer suddenly gives way like a logjam and unleashes a cascade of progress deep into other related areas of study.

By far the most important realization of all has come not from modern science but from ancient expectations about how to tell the Creation story. All things have emerged from simpler forms. They also enter existence in their own season. Scientific stories have, likewise, searched for ancient forms of matter and the rules that govern them. The details, woven into the modern Creation story, have been hard won. They have involved hundreds of thousands of scientists working in sometimes far-flung and seemingly unrelated areas of research. Hundreds of billions of dollars of research funding have gone into refining the basic physical principles, into testing and retesting them for consistency at increasing decimal precision. Dozens of Nobel Prizes have been awarded for key experimental discoveries, which then get woven into an intricate tapestry unlike any body of knowledge ever assembled by humankind in 10,000 years. The miracle of the new story is that it fits so well into the rough outlines of the old revered ones. Again there seems to be some guiding hand at work, always leading us in the same direction in the telling of the story. This time we see how the story is written, not on faded parchment but in the very rocks beneath our feet, in the countless stars above our heads. It is also, as we have begun to learn, written largely in the Void. The most awe-inspiring part of any Creation story is its explanation of how the world emerged from the Void, from a sterile darkness. It is a mysterious process, but one that has been known to science for nearly sixty years. The discovery should have been a historic moment for humanity. It should have caused many people to write inspiring, lofty essays exalting the magnitude of the accomplishment. Instead, it got lost in the technical minutia of an exotic science.

Sometime in the early 1930s, the first trace of matter was torn from the Void by artificial means. Like the world going about its usual business in Bethlehem in 4 B.C., scientists didn't even realize what had happened. No one really knows who reached this frontier first and laid claim to it. Perhaps it was at the General Electric plant in Pittsfield, Massachusetts, where engineers blasted power-line insulators with electrical surges reaching 6 million volts. Perhaps it was in a German laboratory where physicists were experimenting with 1,000-ampere pulses of electrons that reached over 2 million volts. Or perhaps it was at U.C. Berkeley, where a new machine no bigger than an automobile tire passed the 1-million-volt mark. Whatever the event may have been, ghostly blue glows of light slashed many yards into the atmosphere, signaling the birth of new matter. No one was expecting that matter could be formed through human intervention, so no one was looking for it or its telltale traces. There was no concept of the Void as a reservoir of matter or of physicists as demigods able to create substance out of nothingness. It was a miracle lost in time, concealed by the day-to-day minutia of industrial research, but fortunately it was not a miracle whose significance was irretrievably missed. Within twenty years, more powerful machines were built and put into service. This time, physicists wanted to mimic the awesome power of cosmic rays, which rained down upon the Earth from the depths of space. Instead, what they got was a ringside seat for a cosmic miracle restaged on the human scale.

It was April 29, 1949. The Berkeley Synchrotron had been tuned up by Edwin McMillan and his partners so that its invisible and deadly beam would focus on a specific spot in space, with just the right energy. With the throw of a switch, a powerful beam of electrons flashed through the machine. The beam plunged through a platinum plate and exited as a flurry of X-rays, penetrating deeply into a block of carbon. There, as X-rays looped past millions of carbon nuclei, each X-ray photon vanished and spawned a single pair of particles. Photographs recorded the hailstorm of new matter born from the seed of light. Over 20 million particles each hour flashed into existence out of the cloak of nothingness. It was an ugly event, in a laboratory crowded with boxes, desks, and half-eaten sandwiches. There was no music, only the relentless droning of vacuum pumps and machinery. Electrical cables snaked across the floor, which itself had not been swept in weeks. This is hardly the setting for a miracle, but perhaps no more impossible to imagine than a baby born in a manger.

Many more thresholds were crossed in the next few years, as other particles never seen before briefly made their appearance before flashing out of existence in a burst of light. They were all odd particles, these new forms of matter, each destined to live a scant trillionth of a second or less before fading away. But then on September 21, 1955, the first proton-antiproton pairs tumbled into existence at Berkeley's new Bevatron machine. In time, physicists learned how to separate the fleeting pairs of particles to create a new form of matter never intended by nature: antihydrogen. Entire laboratories soon became meson factories and quark factories spewing out millions of particles every year for physicists to study at their leisure, ripping them out of the Void in an impassioned quest to understand not the Void but the particle, as shown in Figure 10.2. We are no longer talking about humans occasionally creating new matter not really knowing what they were doing. By the 1960s and in the decades to follow, creating matter out of the Void literally became a billion-dollar-a-year industry worldwide.

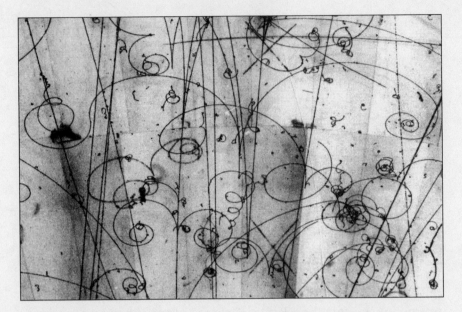

FIGURE 10.2   Invisible particles can be pulled out of the Void. Caught in powerful magnetic fields, they spiral about and are lost amid the hubris of other already existing and familiar particles. (Courtesy CERN/Gargamelle)

Was anyone excited about what had just been accomplished? Scientists were certainly overjoyed when the proton was coaxed out of its hiding place in the Void; however, it was an excitement that celebrated the physics more than the metaphysics. A Nobel Prize was awarded for the antiproton discovery in 1959, though not for the simple but profound, and almost spiritual, act of actually creating it. There were no long speeches about physicists stealing the power of the gods as Prometheus had stolen fire from them ages before. Physicists, ever the careful chroniclers, reported how they created new matter, mentioning only the details of the act itself, never its far-reaching emotional significance. The accomplishments were often couched in the usual self-conscious words of science: "The results prove conclusively that mesons are produced and that they are not the result of secondary heavy particles and most probably not the result of secondary electrons (McMillan, 1949)."

Since the 1930s, physicists have brought into existence trillions of bits of matter that never before existed in the universe. Proton number 1,863,272,195 was never included in the original census of Creation 15 billion years ago. Cosmic rays and the cores of hot stars can also drag particles out of the Void, but what humans had done was unique. The experiment relied on a conscious, premeditated effort to set up conditions on a planetary surface that can mimic what Nature does elsewhere—even at the beginning of time. It is one of those feats in human history that should have been granted far more respect than it ever garnered. It should have rivaled the Moon landing or the discovery of DNA in historical significance. Instead, it has been lost among the incidental discoveries of science that go unnoted. Yet it is *the* event that raises humans to the stature of the gods we have exalted over the millennia.

The countless trillions of particles of matter that humans have brought into existence during the last fifty years are, of course, less than a drop in the bucket compared to the 100,000 trillion trillion trillion trillion trillion particles that exist in the stars and galaxies out to the visible limit of our universe, the product of the Creation event itself. This numerical disadvantage really isn't the point at all. By the act of creating a *single* one of these particles out of nothing, let alone repeating this act trillions of times, humans have demonstrated something magnificent and wondrous. What we do now happens *because* of the way the physical world is *already* put together. Given the properties of the Void as we understand them, God may not have had any choice in bringing forth matter from the Void by the time He got to this point in creating the physical world. Once God created the Void, all He needed to do was say, "Let there be light." The energy of these words reverberating throughout space was enough to excite the Void and bring the world into being. Today, we recreate a small part of this miracle and discover there is no limit to the kinds of particles that can be brought into existence so long as they mesh with Nature's preestablished patterns of what particles can exist, which conforms to a pattern and a dictate that is itself somehow coded in the Void. So the ancient idea of a primordial Void bringing forth matter now seems to mesh with modern scientific understanding, though the ancients expressed the miracle in far lovelier prose. There is, however, much more to the scientific story, and it is this next part that literally lights the world.

For reasons we have yet to fully understand, the Void will not let us create only matter. Antimatter must also flow out of the darkness in equal measure. Yet as we look around us, we see only the one kind of substance. There are only stars and galaxies made from matter, and they are not mingled with their antimatter twins. Where did the other kind of matter go? In the beginning, we see how matter emerged from the Void but also see antimatter flooding Creation. When they meet, they vanish in a blaze of light. Science can make no argument with "Let there be light" as one of the most important ingredients of the primeval universe. The echoes of this ancient light are everywhere around us as a reminder of the ancient conflict between good matter and bad antimatter. But it wasn't all simply endless annihilation. When matter and antimatter decay, matter has a small upper hand. We don't know why, but the arithmetic is always the same. If you had 30 million particles of matter and 30 million particles of antimatter, these individual obese particles would themselves each decay into almost equal amounts of matter and antimatter, which would then annihilate. There would also be about one matter particle left over in the end. In the ensuing annihilation, the 30 million pairs of particles would find themselves and become pure light, leaving behind one particle of matter. We can measure the likelihood of the miracle that is matter, and we find it is almost literally one in a million. Thus, from the Void, the era of light was born. Mixed in with the incomprehensibly bright light would be a trace of matter riding the light like dust trapped in a sunbeam, the sole survivor of a devastating moment in cosmic history. The universe grew in size, cooled, and the light faded into the growing intergalactic darkness over the eons. But that tiny leftover trace of matter survived to form planets, stars, galaxies, and us. Light still envelops us as a permanent reminder of the miracle of our origins. The ancient light is a reminder of how narrowly we creatures of matter managed to avoid complete nonexistence.

Thus, the wellspring of the matter we see around us today was the Void in equal measure throughout all space. Light replaced darkness, and from that came the first traces of matter floating in the newborn universe. This all happened before the universe celebrated its first day of existence, in a murky world at the borderland of science fact and science speculation. And now the child in us steps in with a new question: "Where did the energy come from that shook the Void?" What provides the energy to pull particles out of the Void if there is nothing around to kick with? Humans can create matter from the Void in their laboratories, but they first have to use matter (machines) to supply the kick in just the right way. How did an unthinking cosmos manage to kick itself? The answer to this paradox of producing matter without matter takes us back to our discussion about gravity and how its field operates at the quantum level. The Void exists within the even greater arena of the gravitational field. When this field gets deformed or warped, it causes gravity. If these forces are strong enough, *they* can supply the energy needed to pull particles out of the Void. Can anyone prove this actually happened? Have humans ever warped a gravitational field and caused matter to spew forth? No, but by a mixture of provocative experimentation and logical deduction, physicists can see how insubstantial fields can supply this energy without needing matter as a midwife.

If you ask a physicist what it might take to make the Void bring forth matter, a popular answer might be to simply make a field so intense and unstable in time that quantum uncertainty does the rest of the work for you. This is the same quantum uncertainty that lights the stars and makes your laser pointer work. Physicists first discovered how to do this by using such intense beams of electromagnetic energy as X-rays or gamma rays. But there is also a second way that works. If you collide two heavy nuclei together, the protons of the new nucleus formed this way create an electric field around them so intense the Void becomes as unstable as a rain cloud. Physicists Jack Greenberg at Yale and Walter Greiner at Johann Wolfgang Goethe University did this in 1982 with uranium atoms. For just a few moments, the Void released a few droplets of quantum rain. With nuclei having more than 173 protons, a veritable cloudburst of particles will flow into our universe as it would through a microscopic door left momentarily ajar. The dense atomic nucleus is momentarily surrounded by a luminous halo of destruction as pairs of particles wrenched out of the Void find themselves and vanish. Can gravity do the same thing? Physicists think so. In 1975, Stephen Hawking made us see that even black holes can warp space so badly that their gravity alone can pull matter out of the Void: The tremendous gravitational forces would pull apart the virtual particles that come and go invisibly in the outskirts of a black hole and prevent their annihilation back into the Void. The black hole would glow imperceptibly with the emission of matter that was never there to start with.

We also know the Void is not empty. There seems to be a field within it, a partner to the gravitational field like the meat inside the walnut, which today seems to be causing the expansion of the universe to accelerate. Astronomers don't really know what it is. Physicists have never seen anything like it in their laboratories, either, although some physicists already have a name for it just in case it is found. They call it "quintessence" or "dark energy." The only other field even remotely similar to this one is the mysterious Higgs field, which is supposed to cause things to have mass. These hidden fields, which control so much of the current shape of the universe, have no sources and do not require matter to bring them into existence. They are already present within the warp and woof of the gravitational field. Some physicists are convinced that these hidden fields may have been the source of the energy that created matter from the Void in the first place. Empty space can also cause a force on its own without matter needed as a source. Like the creation of matter, it is one of those discoveries that should have been shouted from the rooftops but never was.

On January 6, 1997, physicist Steve Lamoreaux at the University of Washington took two tiny plates of quartz about the size of a quarter, each covered with a thin layer of silver, and brought them together until they were closer than the width of a human hair or a bacterium. If the Void were sterile, there would be no effect as the minuscule gap closed. Instead, the plates registered an increasing tug that grew in strength as they approached each other. This was no mysterious force. Incredibly enough, it was the Void itself pushing the plates together. A whole new dimension of the Void became a physical reality. It was a peculiar phenomenon that had been predicted fifty years earlier by the Dutch physicist Hendrik Casimir (1909–2000), but it took all that time for physicists to develop the technology needed to produce it and to attempt the tedious experiment.

The Void of Genesis and Maori legend has now made full contact with the modern story of Creation, recast in scientific terms. The invisible, empty Void can create the matter upon which all of the physical world is constituted. It can also create forces that move matter and that in some instances do not even need matter to provide their source. Some physicists and astronomers have become convinced that under the right conditions, the Void can actually create entire universes like ours from its own mysterious fabric. When hidden fields are added to the Void, they cause a tremendous pressure, which completely overwhelms gravity. The universe expands so briskly that it dilated a billion billion times its original size in a trillion-trillion trillionths of a second. With each passing moment, the energy stored in the Void increases until even the Void can no longer contain it. In a spectacular burst of light, matter flashes into existence as matter-antimatter pairs. In the ensuing orgy of annihilation, both the cosmic fireball light and the trace of matter we now see were the only survivors.

The scientific story of Creation might seem compelling in that we may have managed to explain the source of matter, but the Void itself remains mysterious: Where did *it* come from? All we have to do is take one more step back in this story and a turbulent gravitational field confronts us—space and time contorted beyond recognition. But there is still a physical system there, existing in some meaningful way. It still exists on this side of rational thought, but its hold is beginning to weaken. Soon the entire story will slip away from us. We are at the earliest stage anyone can ever hope to connect with observations made today. The matter in our world could have emerged from the Void as the Void was contorted by gravitational stresses driven by hidden fields in an ancient age. Since the Void is just another name for the gravitational field of the universe, and of the plenum of space and time, matter's origins now disappear into the workings of space and time, themselves bound within an incomprehensible Gordian knot of causation without cause, and time without time. Since the gravitational field may be the product of strings that can change their properties, at last a connection can be made between gravity and our own existence. At its earliest definable moments, space-time was a miasma of string loops in the Void. Then this collection became unstable in some mysterious way like water turning suddenly into ice. Its contortions caused some of the strings to pick up energy and become real particles—but only an insignificant few. These few, however, would later become the matter out of which our universe with all its stars and galaxies could be fashioned.

A poet whose name I don't recall once wrote about a freeway that turned into a street, then a hiking trail, then a deer path. It finally ran up the side of a tree, disappearing into a squirrel's nest. We find ourselves in a similar situation as we try to complete the new story of Creation. The universe, too, has undergone many physical changes, but now we have followed its transformations up the trunk of the tree, into a knothole in space-time. There is no conceivable way we could ever verify this part of the story. All we can do is what science has done in the past: look at pieces of the story to see if all the pieces work. We can also try to look for failures in the logic and find better models to work with to create a new logical picture. We can try to create better stories that let the dreamer continue to dream without sensing the descent of the story into chaos.

This chapter began with a discussion about phantom worlds that dissolve away once the sleeper wakes. These dreams seem to follow a logical sequence of events, but they eventually deteriorate into a confusion of unrelated episodes in time and space. Science helps us craft better stories, but eventually they, too, become frayed at their edges. What happens when the dream and the dreamer become one? When matter blends completely with field? What story can be told when time itself abandons us like a thread coming to its end? The primordial waters rise up and absorb the Divine Spirit gliding across them. This is the new paradox we now face, because with superstrings and space-time, we have dissolved the physical world into a story of geometric patterns woven into the Void. To do this, we have had to deny time itself as the final organizer of the story, thus the material storyteller disappears into the fabric of the story. There is no outside point of view, or platform, upon which to view the landscape.

All human beings can be traced through all their life stages by the continuity of their unique DNA. It is a timeline coded in each cell, which shapes a human's outward form from year to year. We can trace this chemical fingerprint that makes each of us different from our parents or any other being back through countless stages of transformation to the very instant a sperm and an egg united, dissolving into permanent genetic union. But what happened before the moment of conception? The essence that is you, your DNA, and your chemical timeline dissolves away into clouds of probability, which become more diffuse with every passing generation. By the time you reach your great-grandparents, the odds that the DNA you now have could have existed from among all the possible unions becomes nearly as improbable as the one-in-a-million miracle of matter surviving the Age of Light. The language we use to tell your final story of "origins" must take on a different cast, a different language. There is no longer a "you"—there is only a shimmering cloud of probability and uncertainty defining your physical essence, trapped between time and un-time. It is the Maoris' "boiling maelstrom out of which emerged space, time and matter," the Samhykas' "space condensing out of the Void."

The elemental ingredients that combine to create the fabric of our physical world may live in an eleven-dimensional world. How do we find our way from there to the world around us? Perhaps, some say, the birth of our universe was like the formation of ice from water. Four of these dimensions grew enormous as the universe cooled and inflated in size from its load of hidden fields, a disembodied Casimir force within the Void driving some of the dimensions of space-time to ever-increasing size. Before then, say Stephen Hawking and countless string theorists, space-time could have existed in a more primitive state where time itself was simply another direction in space. For the first time in human history, the story of Creation has been robbed of the hegemony of time. We have arrived at the beginning and can go no further. The story of causation within the framework of space and time has dissolved into a meaningless semantic discussion of the meaning of "is."

We are creatures of time, and we can scarcely imagine how to craft a story that at some point doesn't involve timelike words. The chief dilemma of all the ones we face in continuing such a story is that when you eliminate time, nothing ever changes. The universe, or its ancient seed, balances for a timeless eternity between action and inaction. Even Einstein's general relativity insists that at the big bang, time and space were created alongside matter. The problem with thinking about this event is that humans insist on using time as an organizing element in the story. Our language is bloated with words that imply some kind of time ordering. Our modern language still leads us astray in authoring a story, because even the concept of a story becomes meaningless when there is no longer any time in which to order events. In fact, without time, even the concept of a "story" becomes a bad vehicle for expressing Creation. When the description is stripped of all its verbs, the physical world dissolves into myriad disconnected states, each existing in a timeless heap of pure being, waiting to be selected so that one comes before the other. Time is the most obvious element of the universe, but we have no idea why or how it came into being. Where was this "A before B" law written into the Void? Even in string theory, there doesn't seem to be much of an answer for why time exists. All ancient stories of Creation mention a state when nothing happened, but that condition was then replaced by some decision or transformation that started the whole story. There must have been an even more ancient command that ancient Creation stories never dwell upon—"Let there be Time"—but we have no idea how the command was invoked. Our minds immediately imagine an instant before the command was uttered, and in reaching for that moment, we are forced to conceive of it in complete contradiction to the nonexistence of time.

We have looked at the story of Creation in its earliest moments to appreciate anew the miracle of our origins in modern terms, but what of the universe's present or its future—and our own? Is there a bigger picture we are missing by dwelling only upon the first instants of existence? The creation of the universe may not only have occurred in the deepest history of our own universe. Some cosmologists have taken an even bigger view of what it all could mean. They ask us to no longer see Creation through the chauvinistic eyes of life in a particular cosmos composed of a particular set of fields and dimensions. Rather than searching for a specific theory that uniquely leads to the particles and forces in our universe, suppose that all pos-

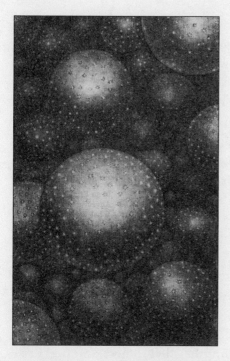

FIGURE 10.3
Many ideas have been followed to describe cosmogenesis, but none seem as audacious as the speculations of modern physicists, though even in their explanations, we can still see glimpses of older stories. Thomas Wright's eighteenth-century cosmos of bubble universes is not unlike the more modern quantum cosmologies of the twenty-first century. (Thomas Wright, *An Original Theory of the Universe*, 1750)

sible "theories of everything" come true somewhere, both those that are logically consistent and those with subtle internal flaws. Some of these might become "big" universes, of which we are just one possibility. Some of the "failed" universes might have existed for only a microsecond, or for a few million years, before disappearing back into the quantum foam of the Void. Even M-theory asks us to think of phantom brane universes floating out there just around the corner from us in hyperspace, like the bubble universes in Figure 10.3 imagined by Thomas Wright.

It could also happen that as the small part of the universe we can see relentlessly expands trillions of years from now, the suburbs of some of these other universes may loom into view. We can only imagine the conflagrations that might take place as different, hotter spaces flood into our own universe with their strange new physics in tow. As physicist André Linde once explained it:

We find ourselves inside a four-dimensional domain with our kind of low-energy physics, not because other kinds of mini-universes are im-

possible, but simply because our kind of life cannot exist in other domains . . . . It is extremely complicated, if not impossible, to construct
a theory in which only one kind of compactification can occur, leading precisely to a four-dimensional, inflationary universe, with the
low-energy particle physics of our own experience. . . . Now it seems
more likely that the universe is an eternally existing, self-producing
entity, that is divided into many mini-universes much larger than our
observable portion, and that the laws of physics, and even the dimensionality of space-time, may be different in each [of them].

Even this exotic idea of a "multiverse" doesn't exhaust the talents of
the modern storyteller. In the Genesis stories of Alan Guth, the frothiness of our universe at the Planck scale may be constantly creating
bubble universes that pinch off from our space-time. Some may even
expand to become universes like our own. A great many universes
could be created from the Void, each disconnected at the largest
scales but in contact at the Planck level via wormholes. There is also a
possibility, according to Lee Smolin, that these universes might resemble our own. Black holes are formed by collapsing stars after the
most massive stars explode as supernovae. What goes on inside them
no one knows, but Stephen Hawking and others imagine that maybe
they cause the creation of another universe elsewhere—perhaps, says
Smolin, even somewhere within the grander arena of Linde's multiverse. Black holes can only form if a certain narrow range of physical
laws is in existence, so universes that find themselves within this narrow range can go on to spawn black holes and more universes of the
same type. Universes like ours with their abundance of exquisite patterns may outnumber ones less able to form black holes that can
punch through the unimaginable Void of the multiverse and spawn
new universes elsewhere and at other times. Science fiction writers
speculate that there might even be civilizations more ancient than
ours, creating artificial black holes to gain numerical advantage for
their own kind in the multiverse.

In the modern story of Creation, gravity is seen as both something
familiar and something dreadfully alien that has been around for a
very long time, even predating the big bang and time itself. Heinz

Pagels offered us this cautionary and thought-provoking comment in one of his entanglements with the earliest moments in the Creation story:

> The nothingness "before" the creation of the universe is the most complete void that we can imagine—no space, time or matter existed. It is a world without place, without duration or eternity, without number . . . yet this unthinkable void converts itself into the plenum of existence—a necessary consequence of physical laws. Where are these laws written into the void? It would seem that even the void is subject to law, a logic that existed prior to time and space.

So what do astronomers and physicists actually believe? The speculations about the pre–big bang state during the last twenty-five years have come as close to a supernatural explanation for the origin of the universe as physics is ever likely to offer. Scientists have also ventured deep into the landscape of what cannot be proved. To many of us, their stories sound more like science fiction than science fact. They are provocative. They are technically well-reasoned, but we have absolutely no idea which of these elements are phantom worlds glimpsed during a waking dream or which are elements of some real story glimpsed through the chinks in a carefully made fence. Although we can dimly see the details of the earliest history literally written in the faded light of the cosmos, one thing we also know about this search for beginnings is that the farther we explore beyond this point, the more expensive it gets. In essence, barring the invention of radically new technology, by the time our comprehension has reached the first plateau at a trillionth of a second after the big bang, we have already come disappointingly close to a practical cosmic censorship about Creation, as resistant to penetration as any horizon ever spawned by black holes. Without new data, no one can put together a reliable, testable theory of space, time, matter, and energy under the conditions needed to understand the details of Creation. As Steven Weinberg reflected in his widely acclaimed 1977 book *The First Three Minutes:* "That which we do now by mathematics [in searching for the underlying unity of nature] was done in the very early universe by heat. Physical phenomena directly exhibited the essential simplicity of nature. But no one was there to see it."

Beyond the first trillionth of a second, some say we must eventually leave the deer path through the underbrush and ascend into the domain of a radically new physical reality fashioned mostly mentally. Here, we will stare upon the face of the true primordial Void and encounter a bewildering bestiary of new phenomena unconstrained by concrete proof or experiment. Yet we also seem destined to meet up with some long-standing paradoxes that will not surrender testable answers: How are the laws of nature written into the fabric of the Void? Was there quantum chaos or perfect symmetry in this initial state? How can experiments be designed to test competing theories? Are there really laws in physics transcending Creation itself? How do we describe laws of nature if our familiar reference points of time and space may themselves be mutable? Perhaps there is a final answer to these questions that is as surprising as it is profound.

Some physicists who don't want to invoke a Divine Creator just yet in the story have adopted something called the "anthropic principle" as at least a temporary refuge from the onslaught of questions without answers. The fact that we are here demands that the path from chaos to reality has to end up with a world with lots and lots of interesting patterns and laws. In many ways, the "why" questions that bedevil us are circular reasoning. Our universe is odd-looking, rich in patterns, and fine-tuned because we are here to wonder about it in the first place. But if this is true, do we really live in the best of all possible worlds, or could the patterns we find have been even better tuned for making life? Suppose that by a better selection of the strength of gravity, or the speed of light, or the mass of the electron, this universe could have become an even more fertile place for sentient life on every world, on every dust mote, perhaps on every star. Could there be universes where the road from slime to sentience is only a few million years, not the terrestrial 4.5 billion? Most of organic evolution on this planet is inefficient and had to wait 3 billion years for free oxygen to appear before it really got going. If sentience is the gateway to the spirit, how much more exalted can a universe become based on matter or on pure field energy? If we really live in a multiverse, there will have to be some universes far more life-nurturing and spiritual than even our own. Perhaps there are universes where consciousness and spirit are also patterns written in the Void, like the delicate patterns of the nebula in Plate 11. Could there be others damned to instant biological decay and constant strife, inimical to the emergence of consciousness under any conditions?

Like our ancestors, we have been drawn to the idea of a primordial Void as the starting point. It is the Alpha and the Omega that contain within them clues to our origins and to our fate in a physical universe. Where our ancestors saw a "boiling darkness," we see fantastic strings of energy vibrating in complex harmonies to create particles and fields. Where they saw desolation, we see the very distinction between time and space slip away into an indistinct state of timeless excitation, perhaps extending into other dimensions. This condition, by any measure, is almost incomprehensibly different from any physical state or system we can now observe in our universe. With only a few minor quibbles, the modern physicist and astronomer would feel comfortable with the prior state described by many of the ancient Creation legends. Why have they come together this way? Is it by divine inspiration, or is it something closer to who we are as a part of the physical world?

> [W]hen we see a new phenomenon we try to fit it into the framework we already have. . . . It's not because Nature is *really* similar; it's because the physicists have only been able to think of the same damn thing, over and over again. (Feynman, *QED*, p. 149)

As we relentlessly think about what patterns might be hidden under all the various descriptions of the primordial state spanning 5,000 years of human inquiry, we sense something that seems to force us to a particular imagery of this state—something lurking silently in the background and, with a subtle hand, steering us toward the same final view of a dynamic Void, regardless of which intellectual starting point we select. And then a second light begins to illumine our understanding of the Void throughout human history. We have all been using the same brains to create the mental images we use and the analogies we explore mentally, and these brains have not changed in thousands of years. A good deal of how we think of the world may be hard-wired and beyond our conscious control. We have at last traced the story of Creation, not into the ultimate depths of a dark Void but into the equally remote depths of how our brains work to extract patterns and meaning out of Nature and fashion stories to explain it all. And so we spin the same kinds of stories, with the same kinds of logical elements. Only their details change.

EPILOGUE

But as I strove to hear more inwardly the music of concrete spirits in countless worlds, I caught echoes not merely of joys unspeakable, but of griefs inconsolable.

—Olaf Stapledon, *The Star Maker*

I'm sitting in my mother-in-law's home in Adams, Massachusetts, where my family and I have spent a lovely Christmas week. A snowstorm busies itself outside, piling up a foot of snow for me to shovel gleefully after breakfast. My daughters and I horsed around with their new Razor scooters and watched Saturday-morning cartoons on TV. My wife, Susan, is chatting with her mother in the kitchen. The rest of the day passes uneventfully. Just a collection of people going about the usual post-Christmas activities that will charge up their psychic batteries for the upcoming weeks at school and work in Maryland. This is a snapshot of life, one that depends very little on the exact way the world is put together, at least from any outward appearances. Our conscious world operates at a level that is light-years above the scale where the physical world goes about its business and where science comes into its own element. We never experience the granularity of the atomic world or the effervescence of light energy. Nature has made sure that we will never experience the Void directly, or any of its hidden ingredients. It has also conditioned each of us to never be conscious of the constant battle we wage against gravity. But like graffiti on a subway wall, this level of reality is there to be appreciated if we know how to look for it, or care to.

239

The biggest scientific success of the twentieth century has been the acquisition of the tools that allow us to look into Nature's tide pools and see clearly what is really going on there. We can gaze deeply into Medusa's face as it is reflected in our sublime mirrors of theory and mathematics. We can observe the cosmic ocean of darkness and see a bewildering world of activity, while at the same time seeing our own familiar reflections in it as physical beings. Logic and experiment have taken us down a road that began with investigation of hard cold matter, then moved on to a plastic, fluid universe where fields are the basic currency and our own substance is mostly an illusion. Is this the end of the journey of discovery or only a way station along some grander road? Whatever the case, it has already opened up some fascinating possibilities for the basis of the universe and for our location within its many patterns. It has also raised a number of questions about the entire framework of reality, making it hard to find an emotional anchor therein.

You have probably gathered from the many anecdotes throughout this book that my own life has hardly been without its curious happenings, from mysterious sounds and ringing doorbells to UFO sightings and alien abduction attempts. My evolution as a scientist began in the complicated and hormone-rich world of my youth, when in grade school I explored both science and the supernatural with equal enthusiasm. For me, science fiction and science fact, linked together by powerful emotional bonds, were not separate universes but just two different ways of looking at the universe. Brain researchers tell us that the adolescent brain is chemically and organizationally different from the adult brain, and I can vouch for this. Because of the stories my mother told me about mysterious events that happened to her family in Sweden, I grew to accept ghosts and extrasensory perception as a real, though elusive, part of the world. In seventh grade, I followed the reports of UFOs in the news and plotted sightings on a map of North America. In high school, I read Edgar Cayce, George Adamski, Immanuel Velikovski, and many other "fringe" authors with the same passion I devoted to Carl Sagan's *Intelligent Life in the Universe.* I learned about Tarot card reading and took part in a spectacularly unsuccessful séance. I read everything I could about "ancient astronauts" who might have built the pyramids or sculpted the figures on the plains of Nazca. My journal entries from these times show my mind as a tumultuous and receptive ground for science and science fiction, myth, and the supernatural. Then when I entered college and studied physics and astronomy, I went through a gradual change. I was still intrigued by UFOs, but my passion for the supernatural waned tremendously. It became harder for me to lose myself in even a well-crafted science fiction story. My post-adolescent brain had finally figured out how to specialize its hemispheres and discount silly ideas over ones that made sense. I learned about the carefully conducted tests for ESP and how the statistics strongly favored randomness. I now understood what such tests meant and implied. They were damning. The next casualty was UFOs, through the overpowering evidence for outright deceit, human misinterpretation of natural phenomena or aircraft, and natural causes. Very steadily, by my senior year in college, all of these ideas became increasingly muted, and I simply stopped thinking about them on any regular basis. By the time reports of people carried off by alien abductors had reached the thousands, it barely registered in my conscious life. What was the big deal? Why couldn't people figure this out for themselves? They were all suffering from harmless "waking dreams."

The bottom line is that my evolution as a scientist completely altered my outlook on the world, expelling virtually all of my childhood ideas about how things work. It also conditioned me to be very suspicious of first impressions, especially if they are accompanied by some overpowering emotional feeling of "rightness." Not so long ago, slavery was considered a "right" thing to do without any question. In some places in the twenty-first century, female genital mutilation is also considered, with great emotional fervor, to be a "right" thing to do. Because I see the many ways that statistics and "eyewitness accounts" can easily mislead, I dismiss nearly all anecdotes by even the most sincere and able observer, even those of other scientists. Because people intentionally, and sometimes unintentionally, deceive others to further their political, philosophical, or religious ideas, I am seldom impressed by any historical story that cannot be supported by archaeological evidence.

It seems that people are never really comfortable with their own history, and much time is spent rewriting it to suit some emotional or political sense of correctness at the moment. Even if some parts of an old story prove archaeologically correct, I never take this as evidence that the entire document is error free. In science, many good ideas confirmed by the data available at one time will be put aside at a later time when new data arrive. I am constantly suspicious of what I believe to be true, for no other reason than a visceral feeling that "it makes sense." The ancient, reptilian part of my brain has an overpowering urge to pay attention to figures of authority. A misfired neuron wants me to believe I just heard the phone ring upstairs. Only my science provides me with a powerful compass to separate truth from fiction, belief from fact. I say this with a certain bitterness, because this ruthlessness comes with a price. I recall endless afternoons lost in a good space novel or in the excitement of placing another pin on the map to mark a new UFO sighting. I also remember the rush of adrenaline upon seeing the mysterious lights over Oakland and hearing the mountain noises in Yosemite. I enjoy hearing about singing sands, and Marfa lights, and frogs falling from the sky. Where do these fit in? They show me that there are still many curious and mysterious things in Nature that deserve appreciation while their origins remain temporarily unknown. But there are many more things that I have had to forever put aside because I no longer have an emotional connection with them. In a strange way, I mourn my lost naïveté. Somehow, the new ideas I encounter as an adult do not have the same powerful draw as the ones I left behind in my youth. Yet there is one area of human exploration that remains almost endlessly fascinating.

A new millennium begins, we enter a new dreamtime in telling the story of how we came to be here, with all the wonderful twists and turns along the way. It is a dynamic story pieced together over centuries of labor, but always under further construction in some remote quarter. Each step along the way has spawned new principles, new technologies, and new possibilities for how we live and work. And we suddenly discover that it is more than a story. It is the alchemist's stone. In creating the stone, we have utterly changed who we are as a civilization and as a species. We have uncovered the subtle workings of the universe and reduced the tyranny of matter to the angelic subtlety of patterns of fields rustling softly in the Void. The Void has been plumbed, and within it we discover the womb of hidden cosmic activity and Creation existing halfway between what is and what might be. We have gleaned the grand spectacle of our universe, its earliest history, and a glimpse of its far destiny: "Even at their physicalistic best [nineteenth-century scientists] left us room for a romantic moment. It was in that century that one prominent scientist described every living thing as 'a melody that sings itself'" (Daniel Robinson, *The Enchanted Machine*).

When you strip away all of its specialized language, the modern story of the Void is different from any other story. It forces us, often against our own sense of who we are, to finally come to terms with the most important elements of our own existence. In the end, it is not what we see that matters or that steers our destinies either as a species or as a living substance. Ninety-seven percent of what you are is a pattern of energy trapped in a gluon field. The three percent of you that tips the bathroom scale each morning is a hint of concrete mass that is a gift from the Higgs field, or something like it, lurking in the Void. The destiny of the entire universe is not controlled by the luminous stars and matter sprinkled throughout space like diamonds on a dark satin cloth. It is controlled by the 97 percent of the dark matter and energy that moves in the Void, unseen except for its feeble gravity. Even who you are may not be under the full control of what you are. Ninety-seven percent of your DNA is in the form of "junk." No one knows what it does or why nature has forced our cells to carry so much of it around. Ninety-seven percent of our DNA, our mass, our cosmos is in a form that is mysterious and unresponsive, and it connects us in an irrevocable way with the dark side of our world—the side that, like Yin and Yang, always accompanies us through life. We see it in the night. We are reminded of it when we enter a dark room. We feel it in our very bones.

We look at the daytime universe and see level after level of things and structures and correlations. We look deep into the nighttime heavens and the Void looks back, with its virtual universe, its shadow matter, its strings and branes. We see ourselves existing only because certain patterns and certain cosmic numbers were balanced just so. Like a pencil on its point, the slightest quantum breeze would have been enough to turn the tide away from life and into the dark abyss of oblivion. Instead, here we are, watching our children grow. What science has now revealed is that Nature is indeed an intricate Gordian knot, a complex webwork of relationships and information, much of which is hidden and invisible, but not unknowable. If we try to break it at any point and eliminate even one of the patterns we discover, the entire work of what we are stands in jeopardy. If the sun doesn't sing or a rainbow doesn't sport its colors in the exact way we see, all the other things we know about in the universe must also change.

On the one extreme, should you now believe that anything is possible, no matter what your wildest imaginings about gremlins, fairies, angels, or past lives? Because the Void trembles with invisible energies, does this mean that our world allows all of these things to exist? Science draws a firm line that says: "No, we live in a rational though complex world ruled by specific laws. Not everything that the mind can imagine is real, no matter how emotionally compelling it is to millions of people." More than that, science gives us a way to decide between the real objective world and the one we create inside our skulls. It lets us distinguish very clearly between the awesome and hidden workings of the invisible Void and the emotion-laden fantasy worlds of channelers, soothsayers, astrologers, and crackpots. Science shows us how our brain hallucinates and sometimes intentionally ignores what the senses tell it. Science explains how we can hear disembodied voices or even ringing doorbells when we are under stress. It shows us how we can see things that are not there. It reveals that our emotions are hard-wired to our perceptions so that whenever we see something unfamiliar or unexplainable in nature, we receive an emotional punch to the solar plexus every time.

The problem is that people don't want to be told that some of their cherished thoughts, feelings, and experiences are not based on more than a misfiring of some neuron, a chemical imbalance, or a chance coincidence. Three million people in the United States suffer from schizophrenia because of a major imbalance in their serotonin levels. No one can reason them into not hearing the voices they hear or not believing the conspiracies they fear. With less severe imbalances, a far wider net can be cast into the human population to reveal temporary moments where phones ring mysteriously, or you hear your name called but no one is there, or you hear telepathic messages from the living or the dead. Still, even for those of us who try to live a completely rational life, the cost is steep. I have lost that simple mysticism I found so stimulating in my youth and now must settle for a more technical sense of wonder, or perhaps insatiable puzzlement, as a mature scientist.

It is popular today to highlight how interconnected we are with Nature, how much a part of the picture we are. Although I can easily accept this idea and have spent nearly this entire book showing you how this seems to be so, I have a hard time working with this idea emotionally. The things "out there" that I find hidden in the information and patterns make me feel more like an exception to the picture than a consistent part of it. There are billions of identical stars out there whose existence is preordained by physics, but there is only one "me." I marvel at the fecundity of Nature in weaving its rich tapestry of detail, level after level, from the cosmic to the quantum. I see the Void dissolve away into fields and strings moving in the night. I can easily understand why rainbows exist and how our eyes and brains see color in them. There is a place for beauty in the physical patterns we uncover. But I can see no role for my consciousness in any of it. There is no place where I can say, "Aha! That's where the pattern of consciousness enters into the equations of reality." With this realization, the whole thing comes crashing down on me. It is exhausting, and as a scientist I simply don't get it. The universe is too big, too old, too poised, too cold. Too much is hidden in places we cannot go. Stars are too far away for us to ever visit them. The night sky is too dark. Why the extremes? And when we die, what was it all for? If the answer is life itself, then why does it have to be conscious? Bacteria do a good job of just "being there," and they will outlast us in the end. Why has the path to consciousness been so badly clouded in random mutations on a minor planet? If consciousness is an exalted feature of a living universe as all of our religions and beliefs tell us, why is it that for 3 billion years, Nature was happy with unconscious things, then in the last 100,000 years decided in a mad dash of activity that consciousness was a worthwhile idea? And then, why do our cells wear out after seventy years or so and consign us to a premature death just at the time when we have begun to figure it all out? The other patterns we see in the world are eternal; only conscious life is temporary. What a waste.

> The worlds opened to our view are graced with wonderful symmetry and uniformity. Learning to know them, to appreciate their many harmonies, is like deepening an acquaintance with some great and meaningful piece of music—surely one of the best things life has to offer. (Frank Wilczek and Betsy Devine, *Longing for the Harmonies*)

Bacteria are immortal, but humans die. And so must the universe, as its own dark Void exacts its final vengeance. It gave us a gift of corporeal existence 15 billion years ago, but in the end the energy debt to the Void must be repaid. As the universe continues its onrush into futurity, there will come a time when all that we have ever known—the lovely stars in the sky, the unimaginably mysterious galaxies, and the very atoms themselves—will pass into oblivion. The great Age of Matter, which lasted for trillions of years, will come to an end. With it will go any record of the life and the consciousness that were once rooted in its fields or matter. It all began with the Epoch of Light—as a great chord sounding in the very fabric of space itself, with the vibrations of individual strings carried into the future and transformed into a spectacular symphony of complex forms and patterns of every possible description. But even the energy of this great cosmic song cannot ring out for all eternity. Vibrations eventually dampen, and what was once a powerful and glorious chorus of creation must in the end diminish until it is but a mere echo in the coming eternal night. Our own deaths will mirror this cosmic death as each of us slides into our personal oblivion, with only a few family anecdotes for future generations to remember us by. In 200 years, not even a picture of you will survive. This is a dreadful story that we latter-day prophets of science have created for the rest of humanity. It cannot end with a hopeful, rising chord if it is a story only of matter and field. It is not a story I would want to tell my children. In 500 more years, we may have a better way to tell it, but not now. All it does is confuse and terrorize. It doesn't even have a happy ending or a comforting word to say about death.

Despite all of its patterns, I sincerely hope there is something unfinished in the universe and its Void. And I wish things were just a little bit different. Wouldn't it be wonderful if the spirits of the deceased could always be here with us in physical form, if ghosts were real and a dramatic part of our world, like a sunset or a tree? If after their deaths, you could still converse with the spirits of your mother and father? There would be no need for channelers or mediums; you could just call their names and the spirit world next door would open up to you for a nice Saturday-afternoon chat. I have often imagined such a world, based on slightly different physical laws, in which what we call a spirit is just another special kind of pattern in the shimmering Void. In this world, telepathy and clairvoyance might be possible. It would not be a world so much different from our own, one would hope, but very likely it would be strange, at least to some degree. We know how this story goes in our world. If you introduce a new pattern into the world or change one that already exists to make room for something new, you have to pay a price. A universe where spirits coexisted with physical beings might not have rainbows. It might not have planets with gossamer rings. The color blue might be absent.

Most people are deeply convinced we already live in such a world. They will tell you that we all have an immortal spirit, that death is a transition. They will tell you with conviction about ancient ghosts wandering the grounds of a churchyard or a castle rampart. People of faith will tell you about saviors, and messiahs, and transubstantiation. All science can tell us is that if spirits do share our world, they have never made a detectable impact in laboratory experiments, which can otherwise sense the invisible activity in the Void. Ghosts and spirits, like the Ether, are ingredients we insist have to be a part of the conscious universe, but searching for them leads only to contradiction. To quote Sir James Jeans, "There is no conspiracy of concealment because there is nothing to conceal." Even the Void, in its cloaked invisibility and with its hidden particles and patterns, leaves a much larger and more obvious footprint for us to find and uncover in an objective way. We do need Heinrich Hertz's "invisible confederates," but they connect with our physical world in what we now see as obvious and measurable ways, despite their invisibility. This is the paradox of modern science and ancient, traditional ways of thinking, and it is the paradox that I seem powerless to resolve.

Writers such as Ursula Goodenough (in *The Sacred Depths of Nature*) have found a way to combine the religious and mysterious elements of their world with the unfolding scientific story. I envy them their convictions. Their writings seem optimistic and hopeful. They are at peace with what they have learned through their scientific scrutiny and religious upbringing. Others such as Stephen Weinberg, in *The First Three Minutes,* seem to lament the pointlessness of it all, disturbed that we can only confront the tragedy of our existence with a heroic search for meaning. The price I have had to pay lies somewhere in between. In traveling nature's many beaches in search of new tide pools to explore, the critical habit of thinking that I developed as a scientist spread its tentacles through my entire waking world. I somehow lost my childlike ability to be awestruck by what I was exploring. I am lost in the technical minutia of Nature with no way to step back and see myself as part of it. As I now learn more about the Void, even this last refuge for my mystical thinking seems to be slipping out of reach. And if I can't find a pattern of consciousness in the world or in a pattern in the Void, then life itself is an accident, and death looms as Nature's end to an accident.

When I am on my deathbed, I will no doubt dissect the experience in the smoldering terror of my last moments of life, which I will probably experience alone in a cold, technology-ridden hospital room as most of us do. There is nothing in my science that will allow me to say good-bye gracefully. It has been a wonderful life, and I will miss my children as they continue where I cannot follow. I am still waiting for a grand event in my life to signify a turning point, a recovery of the sense of mysticism, and, as Einstein put it, an appreciation of the mysterious, which I seem to have lost somewhere. But this event has not yet come, even with the death of my parents and my brother Leonard. They passed into family history with great emotional anguish and despair on my part, magnified by a certainty that I would never see them again on this Earth or anywhere else. They, meanwhile, exited this world with a conviction that some kind of everlasting world awaited them as a doorway opened for them into the light. Surviving siblings mourned their passing but were similarly convinced that the events were only a transition of being, not a termination of it. My thoughts and suspicions were an intrusion, an unwelcome counterpoint to the opinions of others who emphatically believed in a less scientific and more transcendent passage. I feel a great sense of bitterness, sometimes, toward a physical world that withheld just one more message from my mother on the anniversary of her death. Swedish traditions promise this, but the world denies the patterns needed to make it so.

I read Carl Sagan's book *The Demon-Haunted World,* which masterfully summons the arguments he and his family summoned to cope with his own passing, and a cold chill ran down my back. I feel that the kind of science he describes, and to which I as a fellow astronomer am most familiar, has gained us much, but it has also taken away an equally important element of our lives. Something that can help us stare confidently into the darkness and not flinch or despair. Something that can help us deal with our own death and even that of the entire universe when the time comes. Some readers may knowingly call this missing element "God," and there is some part of me that cannot find fault with that opinion. What I do know is this: As I watch my children and gaze lovingly into their eyes, I see my own mortality in a way I have never experienced it before. My science has fully prepared me to meet the Void in all of its technical splendor, but it has failed miserably to provide me with the inner emotional strength needed to face death and darkness, rather than to see this as an opportunity for rebirth. There will be no scaffolding to support me as I slide into my own dark Void. I realize that in the end, no matter what logical universe I will have created mentally, it will be the unreasoning emotional part of me that will color my final moments, and to which I must now devote my remaining years to retrain and comfort. Three thousand years ago, the ancient Egyptians had their own anxieties about darkness and empty places, even to the point of offering a blessing to the traveler to use as an emotional beacon: "May God be between you and harm in all the empty places you must walk." The things that modern science and speculation have turned up in their many walks through the cosmic "empty places" are in some ways at least as disturbing as anything found in ancient descriptions of jackal-headed gods.

But in my personal journey to understand the Void and its hidden wonders, I see a glimmering of hope that in the end I may achieve a more peaceful state of mind. Maybe there is a germ of truth in the notion that "the universe has more of the character of a thought than a machine." In the 1930s, Gilbert Ryle spoke of the human spirit as "the ghost in the machine," dismissing the idea that a spirit cohabited the body. Today we realize that Ryle was wrong. It isn't so much that science has proved there is no "ghost." When you consider that the body is constructed from elemental fields, there is no "machine." It is this realization that gives me some sustainable measure of hope. It is a talisman I carry with me to bind me with the darkness that I most fear: "Individually we come and go but together we contain all of the past, and carry the bright hope of a creative adventurous universe—and not only here but wherever planets have been fertile, wherever a sun shines to bring forth glory" (John Berrill, *You and the Universe*).

On a dark summer's night in July, I stood outside looking up at the vast emptiness between the stars. I saw the familiar dust clouds and nebulae, the faint pinpoints of warmth set against the unimaginable cold of the universe. I tried to make contact with the ancient fears that drove civilizations to sacrifice their own children. I wondered wordlessly about the dark shapes and energies that flowed like rivers between the stars and galaxies, controlling the destiny of our universe. I thought about the virtual particles buzzing around me like gnats, the billions of neutrinos coursing through my body on their way to infinity. As though chanting some ancient mantra, I reminded myself that darkness isn't just the absence of light. It is its own universe of activity, and fields, and things that move about in it. In the end, it will be the Void that survives through the eternities to come, as all the rest of the material world flashes out of existence. In my mind's eye, I saw something else, too. I saw a vast landscape of fields busily curving space-time and steering the motions of matter. Beneath this, I saw ghostlike things that scurried around, knitting the Void together, suspending it like a spider's web above the great abyss of nothingness. I felt the hardness of my body and the ground beneath my feet dissolve away into the invisible gyrations of space-time curvature, in a seamless way reuniting my body with the Void itself. My substance was taken up by the energy of hidden fields, themselves dissolving into the comings and goings of webworks at the foundation of space and time within the Void. Now I had found a way to bind myself to the darkness I had once feared. A part of my mind had at last found the images and patterns that could give meaning to what was once meaningless and unthinkable.

There was no inner voice leading me on in this wordless journey, yet I had the overwhelming feeling, deep within, that all was well. Had some other mute part of my brain found some new pattern in Nature sifted from the information I had filled it with over a lifetime? Had it seen a glimpse of some new possibility, though it could not tell me in words what it had uncovered? If I am lucky in the years to come and recover the gift of mystery, then perhaps I can also hope for a death without fear as my mind takes its last refuge in new possibilities and patterns that no words can describe at this time in my life. For now, as the Sun rises in the East, I am left to marvel at the miracle of life, at how darkness can be cast aside so that I might dance for a brief moment in its light.

GLOSSARY

Dimension   This is a rather Jekyll-and-Hyde term that under normal circumstances is a replacement for "size." Instead of "What is the size of that box?" you could just as easily ask "What are the dimensions of that box?" However, the later usage does seem a bit awkward and overdone in colloquial conversation. When mathematicians, physicists, and astronomers use "dimension," they are not thinking of the measure of a volume size, they mean the number of independent directions or qualities that need to be specified in order to describe something uniquely. The most common of these is location in space, which requires exactly three numbers (three dimensions) to tie something down to a specific location in the universe. Dimensions do not always have to refer to space or spacelike things. Thanks to a reminder from Einstein, we know that you actually need four numbers (three coordinates for space and one for time) to locate something in the actual world. I can tell you I will meet you at the Eiffel Tower for lunch, but unless I give you a date and time, the meeting can never take place. In the quantum world, we can add several other nonspace dimensions to specifying a particle's place in the universe. For example, its spin (0, 1/2, 1, or 2) will tell you what kind of particle is present at a particular location in space and time.

Field   This is one of those oddball words that we know a lot about nonverbally, but when we use it in conversation, it sounds horribly unfamiliar. Our daily newspaper or television weather report presents local and global weather summaries in terms of temperature or pressure maps in full color and with animation. We see numbers assigned to different geographic regions and contour maps in color to show the range of temperature across a continent. This is a direct representation of a temperature field. When we talk about magnetism, we often tack on the term "field" at the end and refer to magnetic fields. Followers of science fiction or the *Star Trek* universe have heard the word "field" used incessantly, and this repetition generally breeds familiarity. A field is nothing more than a way of describing some measurable quality as it changes from place to place in space. For climate and weather across the globe or near your hometown, you can measure air temperature, air pressure, wind speed, water salinity, soil pH, geologic rock type, elevation, magnetic field strength, compass direction, population density, political persuasion, income level, and a host of qualities, and all of these are examples of fields. What we do with this field information depends on what information is being collected, and only in the arena of physics do we hang onto the term as part of the dialogue. Although you may hear about velocity and temperature fields from physicists, you will not hear about political fields and income fields from economists.

Hyperspace This term has been used by science fiction writers for decades and is supposed to mean a collection of dimensions beyond the four that define our space-time (three dimensions of space and one of time). It's not a very accurate or well-defined term, so it isn't used in physics, mathematics, or astronomy. The first use of the term "hyperspace" is difficult to track down, but by 1950, readers of the magazines *Amazing Stories* and *Astounding SF* had already been introduced to it several times. Stories such as Robert Abernathy's "The Ultimate Peril" describe Venusian psychophysicists attacking Earth with hyperspace weapons, and S. M. Tenneshaw's "Who's That Knocking at My Door?" describe a honeymooning couple whose hyperdrive breaks down near a white dwarf star en route to Deneb. In an earlier mention, *Grey Lensman,* published in December 1939, describes a Boskonian attack on the Lensman ship, *Dauntless*, with a weapon that made the crew feel as though they "were being compressed, not as a whole, but atom by atom . . . twisted . . . extruded . . . in an unknowable and non-existent direction." They were no longer in the space they knew, and the speaker speculated that it "wouldn't have surprised me if we'd been clear out of the known universe. Hyperspace is funny that way." The magic of hyperspace set the stage for the favorite mode of space travel, one that completely circumvented the ordinary rules of travel. From this, vast empires and sprawling stories of interstellar adventure were created almost overnight. With few exceptions, the need to explain the details of "hyperdrive" became less intense as the story lines were developed with ever-increasing depth and complexity. With the entire galaxy as a stage, the scale of human science-fiction imagination grew by orders of magnitude. The physical reality of hyperspace travel remains very much an obscure mathematical curiosity because there is no known technology that can change the particular deck of cards dealt to our physical universe. Only quantum particles, exotic strings, or the interiors of black holes have any access to these higher dimensions. Getting there would be a lethal one-way trip. For more about these possibilities and their unfortunate limitations, read Michio Kaku's book *Hyperspace*.

Nothingness Mathematicians call it the "empty set" or the "null set." It is a state that has no characteristics whatsoever. Even the physical vacuum is not a true nothingness because it is regulated by gravitational fields, hidden virtual particles, and a geometric structure that forces it to have specific dimensions. Like "infinity," nothingness is a term that we think we intuitively understand, but in reality we have no way of even expressing it except in terms of the negation of something else as a quality. We cannot think of it on its own terms because there is literally "nothing" we can use to describe it.

Quantum Thanks to car manufacturers' choice of model names and the computer industry, we see this word more often in the news media and on the street than we did twenty years ago. In physics, it simply means the smallest part of a larger system that still retains the descriptive elements of the larger system. An element, for example, can be thought of as the "quantum" of matter. Energy can also come in quantum "packets" and can be exchanged like discrete coins in order to cause matter to give off light or make electrons behave in a certain way within the atom. Occasionally the phrase "taking a quantum leap" is employed to signify a sudden change from one level of operation to another very different one. To make any sense out of the physical world and how it is put together, you will have to reach some level of comfort with this word, because so much of the way the world operates owes its character to its inherently quantum nature.

Space   This is actually a rather mushy word. When we use it, we almost always think of that arena within which astronauts work and stars and planets travel. We speak of "inner space" and "outer space" to define either the microworld of the atom or the private world within our brains and the distant depths of the external objective world outside Earth's atmosphere. We now understand that space is one of the two components of the cosmic gravitational field; the other component is time. Mathematicians also work with different kinds of space, but only a physicist can tell you which of these spaces has a physical meaning.

Space-time   Almost no one knows how to visualize this arena, which is a hybrid form of three-dimensional space woven onto a dimension of time. Circa 1906, Minkowski pioneered this term and found it absolutely central to understanding relativity. Einstein later made it the central concept in describing the gravitational field itself, so we now have to think of space-time as being another name for the cosmic gravitational field. Objects do not move in space-time, they simply exist at all stages of their travel from start to finish along their history, like tracing your vacation travels on a map of the world from your front door to your destination and then back again. Space-time is the "map" that is left after you have drawn the line that chronicles your journey.

Spin   This is a quality of quantum particles and fields that shares a familiar name with things that rotate about an axis, but in fact it has nothing to do with this intuitive phenomenon at all. There are two separate quantum systems of fundamental particles that have specific values of this quality: bosons with integer spin values (0, 1, 2), and fermions with half-integer values (1/2). Bosons include only the particles that cause forces and naturally occur as fieldlike ingredients to nature such as gravity (spin = 2), gluons and W/Z particles (spin = 1), and Higgs fields (spin = 0). Fermions include only particles that seem to appear naturally as the constituents of matter, such as electrons, muons, tauons, neutrinos, and the six different quarks (up, down, strange, charmed, top, and bottom), along with their antiparticles. Spin is one of a set of "quantum numbers" that defines every quantum state for matter and field.

Vacuum   Sometime before the advent of the Pythagorean school in ancient Greece (500 B.C.), the idea of a "vacuum" or "pure empty space" began to appear formally in philosophical discussions. The Atomists believed in its objective existence (450 B.C.), because otherwise atoms could not move. Plato and Aristotle found the existence of a true vacuum logically impossible and "abhorrent" (350 B.C.); and Epicurus of Samos (300 B.C.) later asserted that nothing exists except atoms and the Void. But Aristotle ultimately won the debate, and the impossibility of a true vacuum gained more respect with each passing century, chiefly because of the almost godlike status of Aristotle's pronouncements on the subject of natural philosophy. Following the great intellectual silence of the Dark Ages, the issue of the existence of a vacuum resurfaced between the eleventh and thirteenth centuries. The character of the philosophical discourses, rooted as they were in carefully posed logical "if-then" statements, had barely changed from that of their ancient Greek predecessors. The discussions about the Void led some to the conclusion that if the existence of the Void were logically impossible, not even God with his infinite powers could create such a condition. English philosophers such as Henry of Ghent (1217–1293) soon found a way out of God's dilemma: a Void could exist "accidentally." Henry argued that "the Void had no other existence than an accidental existence, in that the bodies between which it exists are disposed in such a manner that the dimension of a body is capable of being placed between them." God could create a Void, for example, by moving Heaven and Earth, but the Void created by the displacement would be an accidental one since His intention was to move Heaven and Earth and not to create the Void that resulted from this act. Accidents are not covered by natural laws or logical consistency, and therefore God is saved any divine embarrassment. Upon hearing that philosophers were limiting God's power, Bishop Etienne Temprer of Paris condemned all such discourse as heresy in

1277 A.D., and for a while this ended any further discussion of what God could or could not do. What was certain, however, was that at least mere humans could not create a Void. Today, we are surrounded by many technological inventions that require differing degrees of vacuum to make them work, from light bulbs and television picture tubes to the beam line of the Fermilab particle accelerator and NASA's Space Environment Testing Chamber at the Goddard Space Flight Center.

BIBLIOGRAPHY

Abell, George, and Barry Singer. 1981. *Science and the Paranormal: Probing the Existence of the Supernatural.* New York: Charles Scribner's Sons.

Arkani-Hamed, Nima, Savas Dimopoulos, and Georgi Dvali. 2000. "The Universe's Unseen Dimensions." *Scientific American*, August, pp. 62–69.

Asimov, Isaac. 1955. *The End of Eternity.* Greenwich, Conn.: Fawcett Publishing. Science fiction.

Banerjee, Shibaji, Sanjay Ghosh, Sibaji Raha, and Debapriyo Syam. 2000. "Can Cosmic Strangelets Reach the Earth?" *Physical Review Letters*, vol. 85, p. 1384.

Barrow, John. 1994. *The Origin of the Universe.* New York: Basic Books.

Barrow, John, and Joseph Silk. 1983. *The Left Hand of Creation.* New York: Basic Books.

Bartusiak, Marcia. 1988. *Eternity.* New York: Warner Books. Science fiction.

———. 1993. *Through a Universe Darkly.* New York: Avon Books.

Baxter, Stephen. 1993. *Timelike Infinity.* New York: Penguin Books. Science fiction.

Bear, Greg. 1985. *Eon.* New York: Tom Doherty Associates. Science fiction.

Berendzen, Richard, Richard Hart, and Daniel Seeley. 1976. *Man Discovers the Galaxies.* New York: Neale Watson Academic Publications.

Berrill, N. J. 1958. *You and the Universe.* Los Angeles: Science of the Mind Publications.

Blish, James. 1952. *Jack of Eagles.* New York: Avon Books. Science fiction.

259

Bourke, Angela. 2000. *The Burning of Bridget Cleary*. New York: Viking.

Bova, Ben, and A. J. Austin. 1992. *To Save the Sun*. New York: Tom Doherty Associates. Science fiction.

Budge, E. A. Wallis. 1969. *The Gods of the Egyptians*. New York: Dover Publications.

Carr, Bernard. 1994. "Baryonic Dark Matter." *Annual Reviews of Astronomy and Astrophysics*, vol. 32, p. 539.

Chown, Marcus. 1999. "The Fifth Element." *New Scientist*, April 3, p. 29.

_____. 2000. "Before the Big Bang." *New Scientist*, June 3, p. 24.

_____. 2000. "Lost and Found: The Universe's Missing Hydrogen Has Turned Up at Last." *New Scientist*, May 13, p. 18.

_____. 2000 "Shadow Worlds." *New Scientist*, June 17, p. 36.

Clark, Arthur C. 1953. *Childhood's End*. New York: Ballantine Books. Science fiction.

Clark, Arthur, and Gentry Lee. 1989. *Rama II*. New York: Bantam Books. Science fiction.

Close, Frank, and Philip Page. 1998. "Glueballs." *Scientific American*, November, pp. 80–85.

Close, Frank, Michael Martin, and Christine Sutton. 1987. *The Particle Explosion*. New York: Oxford University Press.

Cole, K. C. 2001. *The Hole in the Universe*. New York: Harcourt.

Coleman, Sidney. 1977. "Fate of the False Vacuum; Semiclassical Theory." *Physics Review D*, vol. 15, p. 2929.

Coleman, Sidney, and Frank De Luccia. 1980. "Gravitational Effects on and of Vacuum Decay." *Physics Review D*, vol. 21, p. 3305.

Collins, P.D.B., A. Martin, and E. Squires. 1989. *Particle Physics and Cosmology*. New York: John Wiley and Sons. (Cosmic rays.)

Corliss, William. 1983. *Handbook of Unusual Natural Phenomena*. New York: Anchor Books.

Cowen, R. 2001. "Sounds of the Universe Confirm Big Bang." *Science News*, April 28, p. 261.

Cramer, John. 1991. *Twistor*. New York: Avon Books. Science fiction.

Crease, Robert. 2000. "Case of the Deadly Strangelets." *Physics World*, July, pp. 19–20.

Cytowic, Richard. 1993. *The Man Who Tasted Shapes*. New York: Warner Books.

Darling, David. 1993. *Equations of Eternity*. New York: Hyperion.

Davies, Paul, and John Gribbin. 1992. *The Matter Myth*. New York: Simon and Schuster.

Duff, Michael. 1998. "The Theory Formerly Known as Strings." *Scientific American*, February, pp. 64–69.

Duhem, Pierre. 1985. *Medieval Cosmology*. Chicago: University of Chicago Press.

Eddington, Arthur. 1929. *The Nature of the Physical World*. Cambridge: Cambridge University Press. Quoted in Edwin Schroedinger, *What Is Life?* p. 130. Cambridge, U.K.: Cambridge University Press, 1969.

Folger, Tim. 2000. "From Here to Eternity." *Discover*, December, p. 54. (Interview with Julian Barbour, who discusses why time is a complete illusion.)

Freeman, Wendy. 1998. "The Expansion Rate and Size of the Universe." *Scientific American*, March, pp. 85–89.

Frampton, Peter. 1976. "Vacuum Instability and Higgs Scalar Mass." *Physics Review Letters,* vol. 37, p. 1378.

Gatland, Kenneth, and Derek Dempster. 1959. *The Inhabited Universe*. Greenwich, Conn.: Fawcett World Library.

Glanz, James. 1998. "Exploding Stars Point to a Universal Repulsive Force." *Science,* vol. 279, p. 651.

Glashow, Sheldon. 1991. *The Charm of Physics*. New York: American Institute of Physics Press.

Glashow, Sheldon, and Richard Wilson. 1999. "Taking Serious Risks Seriously." *Nature,* vol. 402, pp. 596–597.

Goldberg, Stanley. 1984. *Understanding Relativity*. Boston: Birkhauser. (On Lodge and his ideas about the Ether).

Gombrich, E. H. 1956. *Art and Illusion*. Princeton: Princeton University Press. (Goltzius woodcut of whale; Dürer sketch of rhinoceros.)

Goodenough, Ursula. 1998. *The Sacred Depths of Nature*. New York: Oxford University Press.

Greene, Brian. 1999. *The Elegant Universe*. New York: Vintage Books.

Grib, A. A. 2000. "Particle Creation in the Early Friedmann Universe and the Origination of Space-Time." *General Relativity and Gravitation,* vol. 32, p. 621.

Gribbin, John. 1993. *In the Beginning*. Boston: Little Brown and Company.

Grinnell, David. 1958. *Edge of Time*. New York: Ace Books. Science fiction.

Gunzig, E., J. Geheniau, and I. Prigogone. 1987. "Entropy and Cosmology." *Nature*, vol. 330, p. 621.

Hamilton, Edmond. 1961. *City at World's End*. Greenwich, Conn.: Fawcett Publishing. Science fiction.

Hertz, Heinrich. 1894. *The Principles of Mechanics Presented in a New Form*. New York: Dover.

Hinton, Charles. 1887. "What Is the Fourth Dimension?" Reprint, *Ladies College Magazine*, vol. 8, September 1883, p. 51. London: Civil Service Printing Company.

Hogan, Craig. 2000. "Cosmic Discord." *Nature*, November 2, p. 47.

Hopper, Judith, and Dick Teresi. 1986. *The Three-Pound Universe*. New York: Macmillan.

Howard, Francis Minturn. 2001. "To an Unborn Archeologist." Poem reprinted in *Archeology*, March–April, p. 12.

Jammer, Max. 1957. *Concepts of Force*. Cambridge, Mass.: Harvard University Press. Reprint, New York: Harper Torchbooks, 1962. (On Descartes's vortex theory, p. 105; on Newton's concept of gravity, pp. 134–139.)

Jeans, Sir James. 1941. *The Mathematical Theory of Electricity and Magnetism*. Cambridge, U.K.: Cambridge University Press.

Kaku, Michio. 1994. *Hyperspace*. Oxford: Oxford University Press.

Kane, Gordon. 1997. "String Theory Is Testable, Even Supertestable." *Physics Today*, February, pp. 40–42.

Kapp, Colin. 1964. *Transfinite Man*. New York: Berkeley Publishing Corporation. Science fiction.

Kauffman, Stuart. 1995. *At Home in the Universe*. New York: Oxford University Press.

Kinoshita, T. 1985. *Shelter Island II*. Edited by R. Jackiw, N. N. Khuri, S. Weinberg, and E. Witten. Cambridge, Mass.: MIT Press.

Lamoreaux, S. 1996. "Demonstration of the Casimir Force in the 0.6 to 6 Micron Range." *Physics Review Letters*, vol. 78, p. 5.

Lederman, Leon. 1993. *The God Particle*. New York: Houghton Mifflin.

Lemonick, Michael. 2001. "How the Universe Will End." *Time*, June 25, pp. 46–48.

Leshan, Lawrence, and Henry Margenau. 1982. *Einstein's Space and Van Gogh's Sky*. New York: Macmillan.

Linde, A. D. 1990. *Inflation and Quantum Cosmology.* New York: Academic Press.

Lindley, David. 1993. *The End of Physics.* New York: Basic Books.

Livingston, M. Stanley. 1969. *Particle Accelerators: A Brief History.* Cambridge, Mass.: Harvard University Press.

Maine, Charles Eric. 1955. *Timeliner.* New York: Bantam Books. Science fiction.

Mallove, Eugene. 1987. *The Quickening Universe.* New York: St. Martin's Press.

McDonough, Thomas. 1987. *The Architects of Hyperspace.* New York: Avon Books. Science fiction.

McMillan, Edwin, Jack Peterson, and R. White. 1949. "Production of Mesons by X-Rays." *Science,* vol. 110, pp. 579–583.

Mitchel, Ormsby. 1868. *Planetary and Stellar Worlds.* New York: Oakley and Mason.

Mithen, Steven. 1996. *The Prehistory of the Mind.* London: Thames and Hudson.

Mohideen, U., and Anushree Roy. 1998. "Precision Measurement of the Casimir Force from 0.1 to 0.9 Microns." *Physics Review Letter,* vol. 81, p. 4549.

Moorcock, Michael. 1965. *The Sundered Worlds.* New York: Paperback Library. Science fiction.

Morton, Oliver. 2000. "CERN Gives Higgs Hunters Extra Month to Collect Data." *Science,* vol. 289, p. 2014.

Mukerjee, Madhusree. 1999. "A Little Big Bang." *Scientific American,* March, p. 60.

Newberg, Andrew, Eugene D'Aquili, and Vince Rause. 2001. *Why God Won't Go Away: Brain Science and the Biology of Belief.* New York: Ballantine Books.

Newton, Isaac. 1675. Letter to Henry Oldenburg, December 13, 1675. Quoted in Max Jammer, *Concepts of Force,* pp. 134–135. Cambridge, Mass. : Harvard University Press.

Odenwald, Sten. 1983. "The Decay of the False Vacuum." *Astronomy,* November, p. 66.

———. 1984. "Does Space Have More Than Three Dimensions?" *Astronomy,* November, p. 67.

_____. 1984. "The Planck Era." *Astronomy,* March, p. 66.

_____. 1987. "Beyond the Big Bang." *Astronomy*, May, p. 90.

_____. 1990. "A Modern Look at the Origin of the Universe." *Zygon,* vol. 25, pp. 25–45.

_____. 1991. "The Cosmological Constant: Seventy Years of an Enigma." *Sky and Telescope*, April, pp. 362–366.

_____. 1993. "The Cosmological Redshift Explained." *Sky and Telescope*, February, pp. 31–35.

_____. 1995. "The Final Frontier of Spacetime." *Sky and Telescope,* December, pp. 24–29.

_____. 1997. "The Big Bang Was NOT an Explosion." *Washington Post*, May 16.

_____. 1998. *The Astronomy Cafe.* New York: W. H. Freeman.

Pagels, Heinz. 1985. *Perfect Symmetry: The Search for the Beginning of Time.* New York: Simon and Schuster. (African song.)

Pais, Abraham. 1982. *Subtle Is the Lord . . . .* Oxford: Oxford University Press.

Parker, B. 1986. *Einstein's Dream.* New York: Plenum Press.

Penrose, Roger. 1989. *The Emperor's New Mind.* New York: Penguin Books.

Pohl, Frederik. 1990. *The World at the End of Time.* New York: Ballantine Books. Science fiction.

Ramachandran, V. S., and Sandra Blakeslee. 1998. *Phantoms in the Brain.* New York: William Morrow.

Raymo, Chet. 1985. *The Soul of the Night.* New Jersey: Prentice-Hall.

Rees, Martin. 1997. *Before the Beginning.* Reading, Mass.: Helix Books.

_____. 1999. "Exploring Our Universe and Others." *Scientific American,* December, pp. 78–83.

Reinhard, Johan. 1999. "At 22,000 Feet: Children of Inca Sacrifice Found Frozen in Time." *National Geographic,* vol. 196, pp. 36–55.

Robinson, Daniel. 1980. *The Enchanted Machine.* New York: Columbia University Press.

Romer, John. 1984. *Ancient Lives: Daily Life in Egypt of the Pharaohs.* New York: Holt, Reinhart and Winston.

Rucker, R. 1982. *Infinity and the Mind.* New York: Bantam Books.

_____. 1984. *The Fourth Dimension.* Boston: Houghton Mifflin.

Sagan, Carl. 1995. *The Demon-Haunted World: Science as a Candle in the Dark*. New York: Random House.

Saunders, Simon, and Harvey Brown. 1991. *The Philosophy of Vacuum*. Oxford: Clarendon Press.

Sawyer, Robert. 1996. *Starplex*. New York: Berkeley Publishing Corporation. Science fiction.

Schwarzschild, Bertram. 2000. "Theorists and Experimenters Seek to Learn Why Gravity Is So Weak." *Physics Today*, September, pp. 22–24.

Segre, Emilio. 1980. *From X-rays to Quarks: Modern Physicists and Their Discoveries*. New York: W. H. Freeman.

Seife, Charles. 2000. "CERN Collider Glimpses Supersymmetry—Maybe." *Science*, vol. 289, p. 227.

Serviss, Garrett P. 1909. *Curiosities of the Sky*, pp. 12–16. New York: Harper and Brothers Publishers.

Simak, Clifford. 1964. *Cosmic Engineers*. New York: Paperback Library. Science fiction.

Sky and Telescope staff. 1998. "Is a High Hubble Constant Hanging On?" *Sky and Telescope*, November, p. 24. (Plot of Hubble Constant from 1920 to present.)

———. 1999. "Distant Stellar Flare Strikes Earth's Ionosphere." *Sky and Telescope*, January, p. 22.

Smolin, Lee. 1997. *The Life of the Cosmos*. Oxford: Oxford University Press.

Smoot, George, and Keay Davidson. 1993. *Wrinkles in Time*. New York: William Morrow.

Struve, Otto, and Velta Zebergs. 1962. *Astronomy of the Twentieth Century*. New York: Macmillan.

Sullivan, William. 1996. *The Secret of the Incas*. New York: Crown Publishing.

Wagner, Walter. 1999. "Black Holes and Brookhaven?" *Scientific American*, July, p. 8.

Watkins, Peter. 1986. *Story of the W and Z*. Cambridge, U.K.: Cambridge University Press.

Weiler, Kurt, and Richard Sramek. 1998. "Supernova and Supernova Remnants." *Annual Reviews of Astronomy and Astrophysics*, vol. 26, p. 312.

Weinberg, Steven. 1992. *Dreams of a Final Theory.* New York: Pantheon Books.

Weiss, P. 2000. "Hunting for Higher Dimensions." *Science News*, February 19, p. 122.

_____. 2000. "Most-Wanted Particle Appears, Perhaps." *Science News*, September 23, p. 196.

_____. 2000. "Seeking the Mother of All Matter." *Science News*, August 26, pp. 136–138.

_____. 2000. "Signs of Mass-Giving Particle Get Stronger." *Science News*, November 4, p. 294.

_____. 2001. "Force from Empty Space Drives a Machine." *Science News*, February 10, p. 86.

Wesley, J. C. 1972. "The Current Status of Lepton 9 Factors." *Reviews of Modern Physics*, vol. 44, p. 250.

Wigner, Eugene. 1960. "The Unreasonable Effectiveness of Mathematics in Natural Science." In *Communications in Pure and Applied Mathematics*, vol. 13, no. 1. New York: John Wiley and Sons.

Wilczek, Frank. 1999. "Black Holes at Brookhaven?" *Scientific American*, July, p. 8.

Wilczek, Frank, and Betsy Devine. 1988. *Longing for the Harmonies*. New York: Penguin Books.

Witten, Edward. 1996. "Reflections on the Fate of Spacetime," *Physics Today*, April, pp. 24–30.

INDEX